Solutions For Sustainability Challenges

AF800453

Wulf-Peter Schmidt

Solutions For Sustainability Challenges

Technical Sustainability Management and Life Cycle Thinking

Springer

Wulf-Peter Schmidt
Center for Advanced Sustainability
Management (CASM)
CBS International Business School
Cologne and Solingen,
Nordrhein-Westfalen, Germany

ISBN 978-3-031-63623-3   ISBN 978-3-031-63624-0 (eBook)
https://doi.org/10.1007/978-3-031-63624-0

© The Editor(s) (if applicable) and The Author(s), under exclusive license to Springer Nature
Switzerland AG 2024

This work is subject to copyright. All rights are solely and exclusively licensed by the Publisher, whether the whole or part of the material is concerned, specifically the rights of translation, reprinting, reuse of illustrations, recitation, broadcasting, reproduction on microfilms or in any other physical way, and transmission or information storage and retrieval, electronic adaptation, computer software, or by similar or dissimilar methodology now known or hereafter developed.
The use of general descriptive names, registered names, trademarks, service marks, etc. in this publication does not imply, even in the absence of a specific statement, that such names are exempt from the relevant protective laws and regulations and therefore free for general use.
The publisher, the authors and the editors are safe to assume that the advice and information in this book are believed to be true and accurate at the date of publication. Neither the publisher nor the authors or the editors give a warranty, expressed or implied, with respect to the material contained herein or for any errors or omissions that may have been made. The publisher remains neutral with regard to jurisdictional claims in published maps and institutional affiliations.

Cover illustration: Author has supplied images: Copyright informations: © Artofinnovation/Generated with AI/Stock.adobe.com

This Springer imprint is published by the registered company Springer Nature Switzerland AG
The registered company address is: Gewerbestrasse 11, 6330 Cham, Switzerland

If disposing of this product, please recycle the paper.

Act only according to that maxim by which you can at the same time will that it should become a universal law.

—Immanuel Kant

For Fabian and Malena Schmidt and all future generations.
With big thanks for their help.

Contents

About the Author

Wulf-Peter Schmidt did his Ph.D. in environmental engineering before working in the automotive industry for more than 26 years. During this time, he acquired experience in various positions relating to sustainability. For the last 10 years, he worked as European Director of Sustainability for a major international automotive vehicle manufacturer (Ford). He also served as Assistant Professor of Royal Melbourne Institute for Technology (RMIT) and is a member of the Editorial Board of the International Journal of Life Cycle Assessment.

He now holds a post at the CBS International School as Endowed Chair of Technical Sustainability Management.

This book summarizes his industry experience and research insights. It aims to support all students in the field of sustainability and provide inspiration for sustainability managers and experts in all sectors and positions. A must-read for Chief Sustainability Officers! This book also serves as the author's call for immediate action on sustainability by encouraging all readers to start implementing sustainability in their respective areas.

Abbreviations

A	Allocation Factor
ADP	Abiotic Resource Depletion Potential
AE	Accumulated Exceedance
AI	Artificial Intelligence
Aka	Also knowns as
AS/NZS	Australian/New Zealand Standards
B	Benefits
B2B	Business-to-Business
BEV	Battery Electric Vehicle
BoP	Base of the Pyramid
C	Complementary Good
CDR	Carbon Dioxide Removal
CED	Cumulative Energy Demand
CFC	Chlorofluorocarbons
CFF	Circular Footprint Formula
CFRP	Carbon Fiber Reinforced Polymer Composites
CHP	Combined Heat and Power
CI	Category Indicator (within LCIA)
COP	Conference of Parties
CPI	Corruption Perception Index
CS3D	Corporate Sustainability Due Diligence Directive
CSO	Chief Sustainability Officers
CSR	Corporate Social Responsibility
CSRD	Corporate Sustainability Reporting Directive
CSS	Carbon Capture and Storage
CTUh	Comparative Toxic Unit for humans
D-FMEA	Design Failure Mode Effect Analysis
D	Debts
DA	Data Analytics
DALY	Disability-Adjusted Life Years
DEI	Diversity, Equality, and Inclusion

DfD	Design for Disassembly
DfE	Design for Environment
DfR	Design for Recycling
DfS	Design for Sustainability
DfSB	Design for Sustainable Behavior
DIS	Draft International Standard
DPP	Digital Product Passport
DQA	Data Quality Assessment
DQC	Data Quality Characteristics
DQI	Data Quality Indicators
DSIS	Design for Social Innovation for Sustainability
DtD	Distance-to-target
DU	Dobson Unit
E	Elementary Flow, Energy, Emission
E/MSY	Extinctions per Million Species
EBIT	Earnings Before Interests and Taxes
EDD	Emotionally Durable Design
E-FMEA	Environment Failure Mode Effect Analysis
EFRAG	European Financial Reporting Advisory Group
E-LCA	Environmental Life Cycle Assessment
ELV	End-of-Life Vehicles
EOL	End-of-Life
ERP	Environmentally Responsible Product or Process
ESG	Environment, Social and Governance
ESI	Economic Sustainability Index
ESRS	European Sustainability Reporting Standards
ETS	Emission Trading Scheme
EU	European Union
EV	Electric Vehicle
FAP	Fair Wage Potential
FMEA	Failure Mode Effect Analysis
GAO	General Accounting Office
GBS	Green Bond Standard
GHG	Greenhouses Gas
GLAD	Global Life Cycle Access to Data
GRI	Global Reporting Initiative
Gt	Giga tons
GWP	Global Warming Potential
HANPP	Human Appropriation Net Primary Production
HC	Human Capital
HFC	Hydro Fluorocarbons
HMW	How Might We
HR	Human Resources
HSS	High-Strength Steel

HUI	Health Utility of Income
HVAC	Heating, Ventilation and Air Conditioning
HVO	Hydrogenated Vegetable Oil
IAE	International Energy Agency
IDIS	International Dismantling Information System
IEC	International Electrotechnical Commission
ILCD	International Reference Life Cycle Data System
ILO	International Labor Organization
IMDS	International Material Data System
IoT	Internet of Things
IPCC	Intergovernmental Panel on Climate Change
ISC	Is-Should not be-Could be
ISO	International Standards Organization
JRC	Joint Research Center
KPI	Key Performance Indicator
LAES	Liquid Air Energy Storage
LCA	Life Cycle Assessment
LCC	Life Cycle Costing
LCI	Life Cycle Inventory
LCIA	Life Cycle Impact Assessment
LCM	Life Cycle Management
LCOE	Levelized Costs of generating Electricity
LCSA	Life Cycle Sustainability Assessment
LFI	Linear Flow Index
LOHAS	Lifestyle of Health and Sustainability
LOVOS	Lifestyle of Voluntary Simplicity
MCI	Material Circularity Indicator
MCRM	Multi-Criteria-Requirement Matrix
MDS	Material Design for Sustainability
MET	Materials, Energy, Toxic emissions
MIPS	Material Intensity Per Service Unit
MJ	Mega Joule
MOSAS	Molten Salt Energy Storage
MSCI	Morgan Stanley Capital International
NC	Natural Capital
NDC	Nationally Determined Contributions
NGO	Non-Governmental Organization
NMP	N-Methyl-2-Pyrrolidone
NMVOC	Non-Methan Volatile Organic Compound
NPP	Net Primary Production
NPV	Net Present Value
ODP	Ozone Depletion Potential
OECD	Organization for Economic Cooperation and Development
OEF	Organization Environmental Footprint

OEM	Original Equipment Manufacturer
O-LCA	Organizational Life Cycle Assessment
ORC	Organic Rankine Cycle
PANEL	Participation, Accountability, Non-Discrimination, and Equality, Empowerment, and Legality
PCR	Post-Consumer Recyclates
PD	Product Development
PEF	Product Environmental Footprint
P-FMEA	Process Failure Mode Effect Analysis
PIR	Post-Industrial Recyclates
PM	Particulates matter
PPA	Power Purchase Agreements
ppm	parts per million
PROAAI	Plan-Research-Observe-Analyze-Adopt-Improve
PSI	Product Sustainability Index
Q	Quality
R&D	Research and Development
RBA	Responsible Business Alliance
RC	Real Capital
REACH	Registration, Evaluation, Authorization & Restriction of Chemicals
REPM	Rare Earth Permanent Magnets
RoI	Return-on-Invest
RSCR	Revenue Share of Circular Resources
SAE	Society of Automotive Engineers
SAF	Sustainable Aviation Fuels
SAQ	Sustainability Assessment Questionnaires
SBTi	Science-Based Target Initiative
SC	Structural Capital
SCUPPIE	Socially Conscious Upwardly-Mobile People
SCW	Supercritical Water
SDG	Sustainable Development Goals
SELCA	Social and Environmental LCA
SETAC	Society of Environmental Toxicology and Chemistry
SFDR	Sustainable Finance Disclosure Regulation
SHDB	Social Hotspot database
S-LCA	Social Life Cycle Assessment
S-LCM	Sustainable Life Cycle Management
SMART	Stretch, Measurable, Aligned, Realistic, and Time-bound
SO-LCA	Sustainability Organizational LCAs
SSI	Sustainable Society Index
SWOT	Strengths-Weakness-Opportunity-Treats
TCFD	Task Force on Climate-Related Financial Disclosures
TCO	Total Cost of Ownership
Tg	Terra gram

UN	United Nations
UNEP	United Nations Environmental Program
WfH	Work from Home
WGI	Worldwide Governance Indicators

Chapter 1
Introduction: The Global Challenge

Abstract Technical sustainability management must contend with a wide range of challenges, ranging from environmental, social, and economic problems to governance, regulatory, cultural, institutional, political, technical and human issues. Environmental issues that relate to human impacts beyond defined planetary boundaries like climate change or species extinction are discussed. Social issues are part of 8 of the 17 UN Sustainable Development Goals. Economic challenges include net capital inheritance by the next generation and, on an organizational level, transformation requirements for non-sustainable businesses. The chapter summarizes challenges that need to be solved urgently.

Keywords Sustainable development goals · Climate change · Water footprint · Planetary boundaries · Human trafficking · Intergenerational · Intragenerational · Corruption

Sustainable management for the planet must contend with a wide range of challenges. From an environmental perspective, it includes global issues like ozone depletion or global warming that have the potential to significantly impact biodiversity by extinguishing global life on a large scale. Additionally, there are regional issues like acidification, eutrophication, as well as various toxicity impacts that occur on local, regional, or even global scale. It also covers social aspects like poverty, hunger, forced labor, diversity and equity, education, and more. Also, economic aspects like decent work and economic growth are part of it. Many challenges are covered by the 17 United Nations Sustainable Development Goals and their 169 targets for 2030 [1]. Sustainable development is defined as "development that meets the needs of the present without compromising the ability of future generations to meet their own needs" [2]—a definition incorporating a strong element of intergenerational equity and justice. This kind of development is limited by environmental, social (society), and economic constraints and requires good governance [3] (Fig. 1.1).

According to the United Nations, midway between the agreement on the SDGs and their target date of 2030, progress is slow and insufficient for eight out of the 17 SDGs, while for 30% of the SDGs progress has stalled or worsened [4].

© The Author(s), under exclusive license to Springer Nature Switzerland AG 2024

W. Schmidt, *Solutions For Sustainability Challenges*,
https://doi.org/10.1007/978-3-031-63624-0_1

© Wulf-Peter Schmidt 2024. All Rights Reserved.

Fig. 1.1 17 Sustainable Development Goals (SDGs) covering environmental, social and economic, and governance dimensions

> Sustainability is about meeting social, environmental, and economic needs of current and future generations with good governance. UN Sustainable Development Goals provide guidance.

Overcoming the current unsatisfactory progress towards the SDGs requires a comprehensive approach. Therefore, the following chapters offer a fundamental, holistic perspective that analyzes the current status of and future options for improvement in sustainability. They combine overall strategy, governance, and management approaches with specific technical approaches and cover various sectors, stakeholders, and organizations. Combining management and life cycle approaches with technical solutions in a consolidated and coherent approach—aka Technical Sustainability Management—is meant to offer new opportunities. It stands in stark contrast to the more common individual focus on life cycle science, sustainable business strategy or sustainable technologies. Technical Sustainability Management is not (only) aimed at specialized sustainability officers but is meant to provide tools for every function in an organization to improve sustainability.

1.1 Environmental Dimension

Environmental sustainability is based on the integrity, biodiversity, and carrying capacity (or planetary boundaries) of ecosystems. It emphasizes intergenerational aspect of sustainability—maintaining good environmental conditions for future generations (Fig. 1.2).

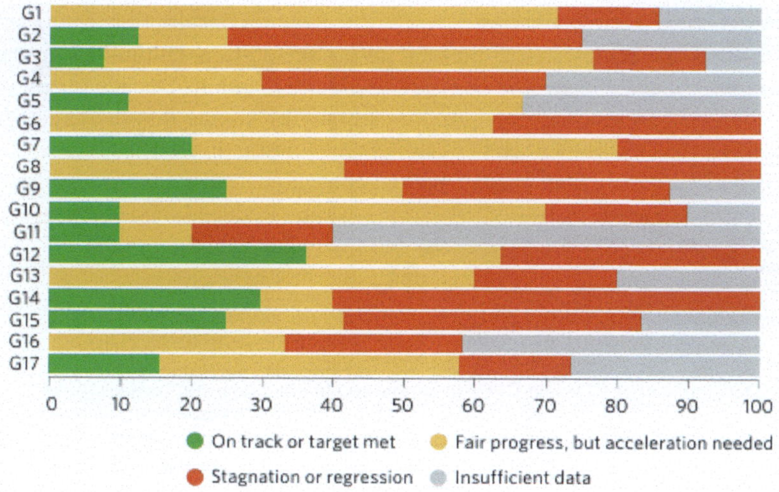

Fig. 1.2 Progress or regress for the 17 UN Sustainable Development Goals for 2030 measured by looking at their respective sub-targets (status: 2023 or latest data) [4]

A good example of regaining ecosystem integrity is the ozone layer. Global efforts of scientists and governments have been and are necessary to achieve global agreements tackling environmental constraints and challenges for the integrity of ecosystems. This has to happen in cooperation with or in opposition to industry. The most effective agreement has been the Montreal Protocol that successfully avoided destruction of the stratospheric ozone layer. This layer is critical for protecting human life against aggressive, carcinogenic ultraviolet light from the sun. The negative impact of chlorofluorocarbons (CFCs) and other substances had first been proposed in 1974, leading to international conferences to tackle the issue of an increasing destruction of the stratospheric ozone layer. The 1981 Vienna Convention first acknowledged the issue while the 1987 Montreal Protocol (enacted 1989) mandated a phase-out of ozone-depleting substances like CFCs [5, 6]. Already a few years later a decrease of CFCs could be observed (see right side of Fig. 1.3). This was a huge success for scientists and concerned governments but it was also possible because substitutes were found by industry. In the beginning, the chemical and refrigerant industry denied any environmental impacts until pre-existing, workable solutions were adopted by some in the industry that embraced the necessary change [5]. This aligns with recent experience that an embracing, "Yes And" position is a helpful catalyst of transformation in industry (see also Sect. 9.2).

In contrast, international agreements to combat Climate Change do not show the same sign of success yet although this issue is explicitly mentioned as UN SDG 13 ("*Climate Action*"), and the impact of CO_2 on climate change was discovered much earlier than the impact of ozone depletion. In fact, the CO_2 emissions continue to increase regardless of all international agreements established so far (see left part of Fig. 1.3).

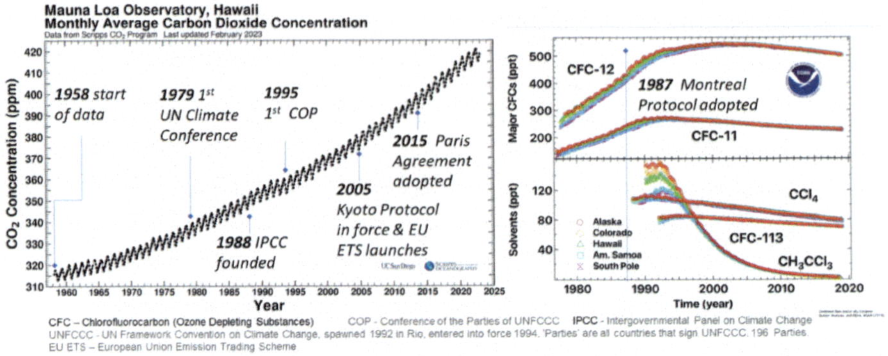

Fig. 1.3 Development of CO_2 and CFCs over time and international conferences [6–9]

Already in 1856, Eunice Foote published a paper in the American Journal of Science and Arts that linked global temperatures with atmospheric CO_2 levels. Independently, John Tydall identified this relationship between CO_2 and global temperatures in 1861 in the Philosophical Transactions of the Royal Society of London. Other scientific breakthroughs—for example, by Svante Arrehnius in 1890s, E. O. Hulbert in 1931, and Guy Callendar in 1938—and finally better technology and more accurate measurements helped to gather sufficient scientific evidence to warn policy-makers from the mid-60s. However, doubts and organized opposition, for instance, by the American Petroleum Institute continued until the end of the last century [10]. Obviously, the necessary changes are more complex than in the case of CFCs. However, there is the hope that with technical solutions getting increasingly available and affordable, industry is starting to support change [11], ultimately leading to measurable positive effects on atmospheric CO_2 concentrations.

Constructive cooperation of governments, industries, and citizens based on scientific guidance is key for success. Ozone layer protection is the role model, but climate change policy lacking.

The key question is whether this will happen fast enough. An analysis of all relevant governmental plans and projections concludes that global coal production will peak in 2030 while global oil and gas production will rise until 2050 [12]. Even best-case forecasts suggest that the target of the Paris Agreement to keep temperatures "well below 2 degrees above pre-industrial levels" and pursue efforts "to limit the temperature increase to 1.5 °C above pre-industrial levels" [13] will not be achieved with current plans (Fig. 1.4). The 6th Assessment Report of the Intergovernmental Panel on Climate Change (IPCC) concludes with "high confidence" that "Human activities, principally through emissions of greenhouse gases, have unequivocally caused global warming, with global surface temperature reaching 1.1 °C above 1850–1900 in 2011–2020… This has already caused widespread impacts and related losses and damages on human systems and altered terrestrial, freshwater and ocean ecosystems worldwide" [14]. The same source concludes with "very high confidence" that climate change has already led to increased mortality rates for humans as well as fauna and flora on land and in the sea. In fact, the impact of increased global temperatures is worse in this IPCC report as shown in Fig. 1.4. Therefore, by summer 2023, 2300 jurisdictions with around 1 billion inhabitants in 40 countries had declared climate emergencies [15]. Even worse, January 2024 marked the first time that the global average surface temperature exceeded 1.5 °C during a 12-month period, according to the Copernicus Climate Change Service. The global mean temperature for the twelve months of February 2023 to January 2024 is the highest on record, at 0.64 °C above the 1991–2020 average and 1.52 °C above the 1850–1900 preindustrial average [16]. Equally February 2024 showed extreme air temperatures around the world. Actually, from June 2023 to February 2024 each month was the warmest on record so far (February 1.77 °C above preindustrial levels) [17] and March 2024 showed unchartered territory in terms of global sea surface temperatures. This does not mean the 1.5 °C limit has already exceeded over a longer period, having also in mind that this exceedance was supported by the temporary El Niño effect.

It is important to note that every tenth of a degree Celsius change in global average temperatures matters and makes a difference. The difference between the Paris climate targets of 1.5 and 2 °C is remarkable. An increase from 1.5 to 2 °C above preindustrial levels increases the population exposed to severe heat at least every 5 years by a factor of 2.6 (to 37% of global population), the species lost by a factor of 2–3, the area of thawing permafrost by 38% and results in almost all coral reefs disappearing [24]. Even worse, the development is not just linear but self-accelerating, in particular when approaching tipping points. A tipping point is a "critical threshold beyond which a system reorganizes, often abruptly and/or irreversibly" [14]. The risk of critical tipping points for climate change is moderate

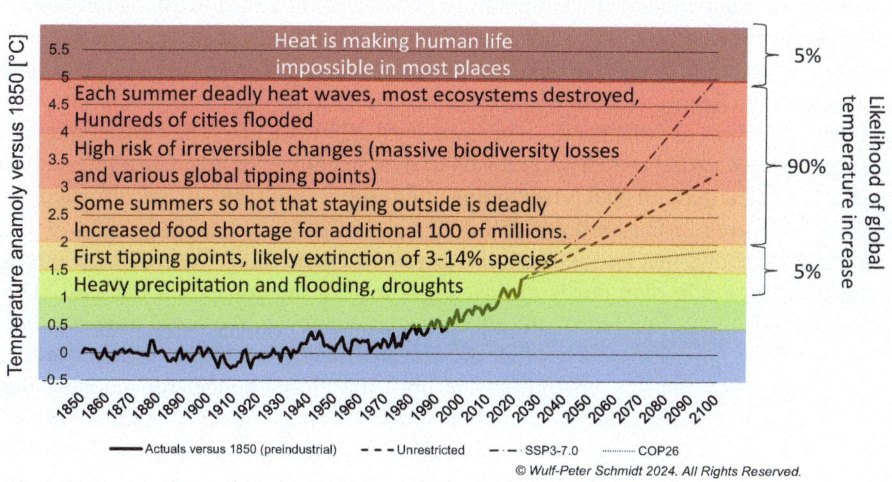

Fig. 1.4 Impact of raising temperature (based on data from [18–23])

between 0.6 and 1.6 °C above preindustrial levels. However, the risk disproportionally increases between 1.6 and 2.6 °C. The larger critical tipping points affected in this temperature range are the Greenland and West Antarctic ice sheets, the slowdown of the Gulf stream, the El Niño–Southern Oscillation, as well as the Southern Ocean role in the global carbon cycle. For example, the loss of ice sheets is likely to happen between 1.5 and 2 °C, leading to sea level increases of around a meter and a change in albedo that reduces the reflection of sun light and therefore further accelerates climate change [14]. This for climate protection negative, self-accelerating feedback loop is typical across several tipping points. The collapse of the Atlantic Ocean's great overturning circulation combined with global warming could cause half of the global area used for growing wheat and corn to be lost. Five major tipping points are already at risk of being crossed imminently due to global warming (the Greenland and West Antarctic ice sheets, warm-water coral reefs, North Atlantic Subpolar Gyre circulation, and permafrost regions) while three more are threatened once the world exceeds 1.5 °C of global warming (boreal forest, mangroves, and seagrass meadows). Currently, more than 25 negative tipping points for climate are known [21].

Each increase of a tenth of a degree in global average temperatures matters and crossing one of the more than 25 negative tipping points can create self-accelerating feedback loops.

Given the drastic global consequences of failing to meet the Paris Agreement, more actions are required in regard to the ongoing climate crisis. In 2018, the remaining CO_2 budget before reaching 1.5° above preindustrial levels was 420 Giga tons (Gt) CO_2 equivalent to achieve this limit with a 67% likelihood or 580 Gt if a 50:50 chance is seen as acceptable. The analogous remaining CO_2 budgets are 800 Gt and 1040 Gt CO_2 equivalent if the temperature increase should be at least "significantly less than 2 degrees Celsius" (1.75°) in line with the Paris climate agreement [25]. Of this budget around 227 Gt CO_2 was already consumed from 2019 to 2022 and the Nationally Determined Contributions (NDCs) for future emissions submitted by countries fall short of what needs to be done until 2030 [14, 26]. The emission levels have already exceeded pre-COVID levels to new negative records. Current policies lead to 3 °C. If at least all NDCs (including conditional ones) plus all net zero pledges are fully delivered 2 °C is still possible [26].

Figure 1.5 shows that late actions typically do not support the Paris climate agreement and any action plan needs to have significant reductions from the start. The earlier significant carbon emission reductions are achieved, the more time remains to tackle those emission sources that are most difficult to address. A life Carbon clock [27] is showing the remaining time until the global budget for reaching the 1.5 C limit is exhausted. The need for front-loaded actions applies to national plans, company plans, as well as personal plans. Similar to the successful example of the historic CFC phase-out, this requires actions by all parties—scientists, regulators and policy makers, industry, citizens/voters, and consumers. This stresses the need for each person and each role on all levels to take responsibility for sustainability with urgency. This will be detailed in the following chapters. Unfortunately, current plans further increase emissions (i.e. even worse than emissions pathway 1 of Fig. 1.5) [12].

> The earlier decisive carbon reduction actions are implemented, the more time is available to find solutions for so far unavoidable emissions while staying within the carbon budget limits.

The same applies to other areas of sustainability. For example, the currently ongoing "sixth mass extinction" triggered by humanity is leading to 100–1000 times faster extinctions of species than would occur naturally. Ten percent of genetic diversity of plants and animals may already have been lost over the past 150 years [28]. This is not just fatal for the affected species. Biodiversity is crucial for the stability of all natural ecosystems and for the functioning of human life-support systems,

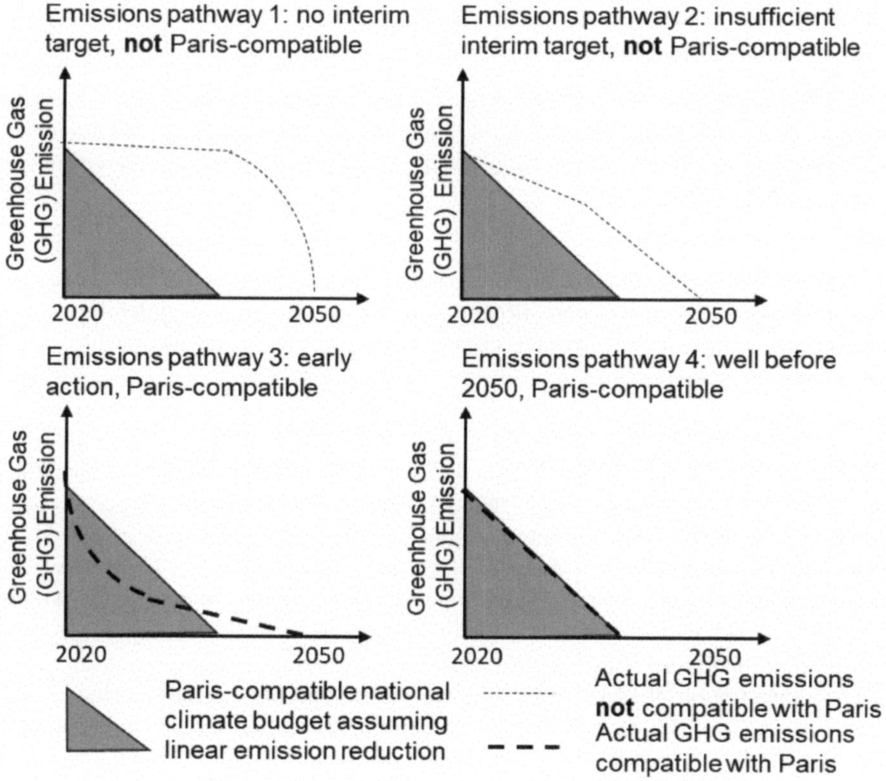

Fig. 1.5 Compliance of general 2050 Greenhouse gas emission reduction pathways [25]

not only in agriculture. It also affects most of the industry and the stability of the global economy as well as societies [29]. Biodiversity is the "variability among living organisms from all sources including, inter alia, terrestrial, marine, and other aquatic ecosystems and the ecological complexes of which they are part. This includes diversity within species, between species and of ecosystems" [30]. The increased loss of biodiversity is driven by a mixture of land use change, overexploitation, climate change, deforestation, fragmentation, and other factors [31]. The impacts of biodiversity losses will last even longer than those of climate change. This was demonstrated in an interim climate change overshoot 2 °C scenario where peak emissions lead to more than 2 °C temperature increase, followed by temperatures returning back to 2 °C after 60 years due to carbon removal. Models suggest risks to biodiversity will occur suddenly when exceeding temperature limits and risks will decrease only gradually and last longer. It will take 100–130 years to get to pre-overshoot levels for some biodiversity aspects while levels will obviously never return to normal for extinct species [32]. In other words, a reliance on carbon removal will be detrimental to biodiversity because it comes too late and creates an interim temperature overshoot. Because of the multitude of risk factors, loss of biodiversity is even more difficult to tackle than climate change. At the same time, an increase in biodiversity can help tackle climate change by offering carbon sinks and resilience. Biodiversity and climate change are closely interlinked [33].

Challenges to and opportunities for biodiversity are more complex and longer lasting than climate change. Biodiversity suffers from climate change but increases climate resilience

Again, global agreements are key. For example, on March 4th, 2023, the "Intergovernmental conference on an international legally binding instrument under the United Nations Convention on the Law of the Sea on the conservation and sustainable use of marine biological diversity of areas beyond national jurisdiction" (BBNJ) agreed to turn 30% of global oceans into protected areas as well as guiding principles like the *polluter-pays principle* and "fair and equitable sharing of benefits" after 19 years of discussions [34]. The UN SDG 15 (Life on Land) mentions biodiversity as one of its sub-targets but does not sufficiently reflect its importance from today's perspective. However, with the Kunming-Montreal Global Biodiversity Framework as agreed on at the 15th meeting of the Conference of Parties (COP) to the UN Convention on Biological Diversity, a similar process to climate change has been initiated with a global framework requiring national biodiversity strategies and action plans as well regular COP meetings [35].

The consumption of fresh, drinkable water is another aspect that has increasing importance to avoid overexploitation of natural resources and keep the global ecosystem in balance. It is also a component of the UN SDG 6 (Clean Water and Sanitation) and SDG 14 (Life Below Water). The trend in water usage is even worse than the trend of carbon emissions (Fig. 1.6).

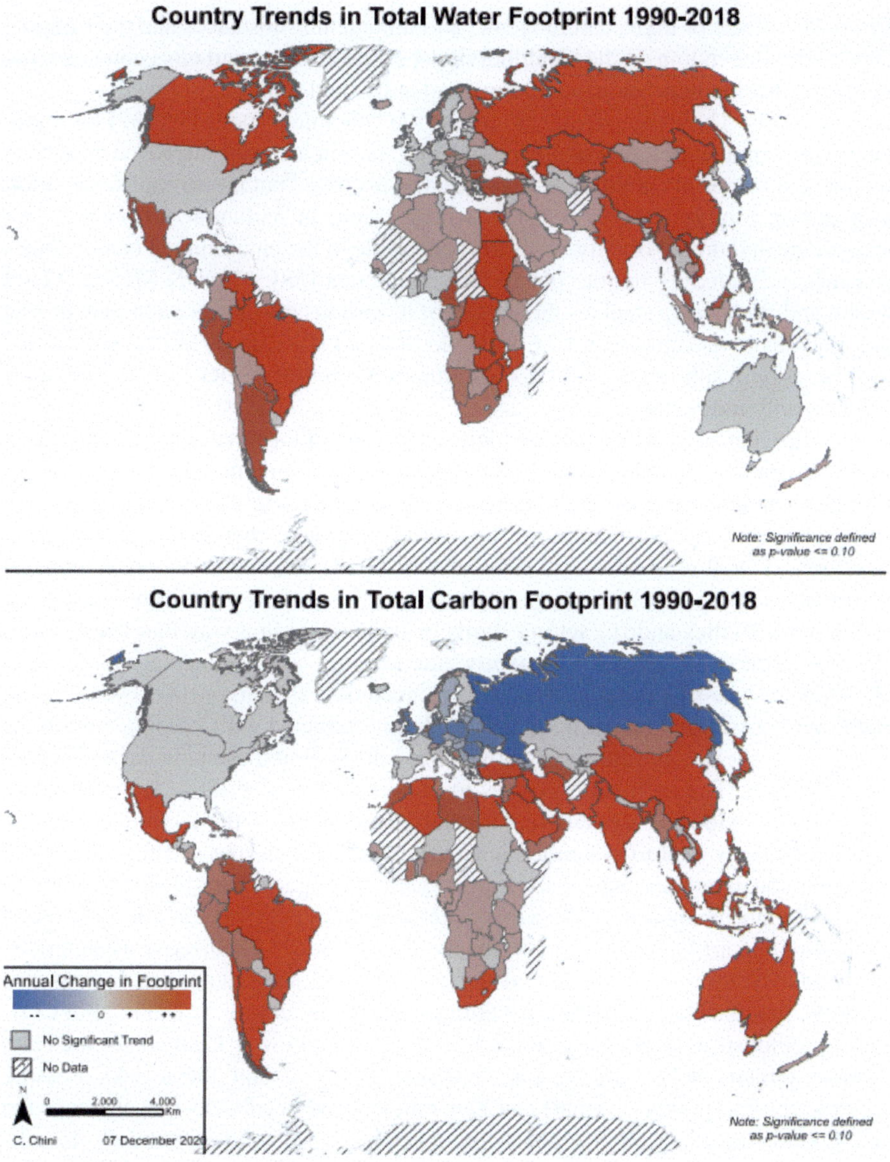

Fig. 1.6 Country trends in total water respectively carbon footprint [36]

This is, in particular, relevant for countries with high water stress that export water-intensive food, materials, or products (like beef, cotton, lithium).

Providing a global picture of other harmful emissions is more complicated than for the above-mentioned aspects of ozone-depleting substances, greenhouse gas emissions, or water consumption. This is mainly because there are countless modes of action and effects of toxic, carcinogenic, and mutagenic substances and their importance varies depending on the local/individual doses, background emission level, the recipient (species, age, health condition), and individual sensitivity as well as historic damage levels. There are attempts to assess the "chemical footprint" [37] but these are affected by the aforementioned complexity and their results suggest a reduction in chemical footprints in Europe. The observed positive "but not strong" correlation with carbon footprint has to be taken with caution. In particular, substances that accumulate and do not or only very slowly degrade in the environment can be found everywhere, including remote places like Antarctica [38]. The UN SDG 3 (Good Health and Well-being) tackles the impact of hazardous chemicals on human deaths and illnesses in its sub-target 3.9 [39] while SDG 14 (Life under Water) covers this issue broadly in sub-target 14.1 (Reduce Marine Pollution). SDG 15 (Life on Land) covers it only indirectly.

A comprehensive and global overview of the state of the environment is provided by the "planetary boundaries framework" that focuses on "nine processes that are critical for maintaining the stability and resilience of Earth's systems as a whole" [40]. Earth system means that all of the Earth's physical, chemical, and biological processes as well as their interactions are modeled. The methodology is human-centric in the sense of (a) focusing on Earth processes that are significantly impacted by human activities and (b) setting limits or boundaries in a way that keeps Earth in a "safe operating space for humanity" i.e. keeping "environmental systems and life-support systems similar to those experienced over the past ~10000 years". The methodology is based on earth science modeling, targeted with keeping conditions of the Holocene, i.e. the current geological epoch. Setting these boundaries is obviously difficult. The methodology tries to identify for each process "a level where the available evidence suggests that further perturbation of the individual process could potentially lead to systemic planetary change by altering and fundamentally reshaping the dynamics and spatiotemporal patterns of geosphere-biosphere interactions and their feedbacks". While this statement implies the methodology is based on state-of-the-art knowledge, it still involves a multitude of subjective assumptions that need to be transparently justified and discussed before drawing conclusions. For example, the planetary boundary for climate change is set at 1 °C (~350 ppm CO_2) above preindustrial levels because the 1.5 °C or well below 2 °C targets of the Paris Climate agreement [13] are seen as "not safe as 1.5 °C and above risks crossing multiple tipping points ... (that) can generate positive feedbacks that increase the likelihood of crossing other climate tipping points i.e. leading potentially to self-perpetuating changes" [41]. The limit of 1 °C is justified by the authors [40] in spite of being above the "observed range through the Holocene up to the Industrial Revolution (for $CO_2 \approx 280$ ppm)" having in mind the "buffering resilience of Earth system… (and recognizing that)… (t)he stability and characteristic range of variability of interglacial Earth system states in Pleistocene paleoclimate and Earth system modeling suggest that the Earth system would likely remain in a stable, Holocene-like state if all boundaries were respected despite them being at least temporarily outside the envelope of Holocene variability". To interpret the results of planetary boundaries, it is important to understand that the boundaries are set on a precautionary principle.

The planetary boundary methodology follows the precautionary principle by targeting conditions equivalent to the current geological epoch (Holocene) for nine indicators.

The biosphere integrity boundaries are set in a way to ensure the "overall dynamic and adaptive character of the biosphere" and keep the change in limits that preserve this character. While the genetic integrity sub-metric of biosphere integrity is easily understandable (see also biodiversity discussion above), the functional integrity requires more discussion. Initially, the planetary boundary method used the Biodiversity Intactness Index but this has been replaced by a new metric that is seen as better linked to the Earth system's functions. The idea is that the net primary production (NPP), measured as photosynthetic energy and material flow available to ecosystems, remains sufficient in light of those energy and material flows diverted to human society—so-called human appropriation NPP (% HANPP). The NPP level increased from preindustrial Holocene base values due to carbon fertilization—a kind of Earth system resilience response dampening the anthropogenic global warming and therefore not available for harvesting. The planetary boundary limit is set at levels typical for the late nineteenth century "a time of considerable acceleration in land use globally with strong impacts on species, already leading to early concerns about the effects of this large-scale land transformation" [40]. This is again a very strict definition of planetary boundaries based on an extended precautionary principle. The intention is to show the overexploitation by harvesting and land use. The authors claim that feeding 10 billion people is possible within the planetary boundaries.

Table 1.1 and Fig. 1.7 provides an overview of all nine planetary boundary indicators and their status. The boundaries are transitional zones where systemic planetary shifts in the Earth system are increasingly likely rather than a sharp, singular threshold. This zone of increasing risk of Earth system losing Holocene-like characteristics is used to assess the status of transgressed boundaries.

Global efforts are required to improve the status towards safe operating conditions while still supporting human needs for a sustainable development. For example, a global circular economy could fulfill people's needs with only 70% of the materials now extracted and used. A circular economy is a system where materials never become waste but is kept in circulation through processes like maintenance, reuse, refurbishment, remanufacture, recycling, and composting and where nature is regenerated. Circular economy is a key tool to bring human activity back towards the safe limits of the planet because material extraction and use is the main reason for a significant part of all emissions linked to human activity and impacts several of the nine areas of planetary boundaries. Unfortunately, only 7.2% of the global economy is circular—based on a declining trend (2018 9.1% was circular). Unleashing 16 transformational circular solutions across four key systems—food systems, the built environment, manufactured goods and consumables, mobility and transport—can reverse the current overshoot of five of the nine key planetary boundaries, thereby

Table 1.1 Planetary boundary indicator [40]

Earth system process	Description	Preindustrial Holocene base value	Planetary boundary	Upper end of zone of increasing risk	Status
Biosphere integrity	Genetic integrity—measured in extinctions per million species and year (E/MSY)	1 E/MSY	<10 E/MSY	100 E/ MSY	>100 E/ MSY
	Functional integrity—measured as % of energy/ material carbon flows diverted to human society (HANPP)	1.9% (2σ variability of preindustrial Holocene mean NPP)	<10% (90% remains for biosphere function)	20%	30%
Climate Change	Atmospheric CO_2 (ppm)	280 ppm	350 ppm	450 ppm	417 ppm
	Anthropogenic radiative force at the top of atmosphere (W/m^2)	0 W/m^2	1 W/m^2	1.5 W/m^2	2.91 W/m^2
Stratospheric ozone depletion	Global average Stratospheric O_3 concentration (Dobson Unit, DU)	290 DU	5% reduction (276 DU)	261 DU	284.6 DU
Ocean acidification	Carbonate ion concentration, average global surface ocean saturation (Ωarag)	3.44 Ωarag	≥80% Ωarag of preindustrial	2.75 Ωarag	2.8 Ωarag
Biogeochemical flows: Phosphate and Nitrogen cycles	Global: Phosphor (P) flow (freshwater → ocean); regional: P flow from fertilizers to erodible soils (Tg P/ year)	0 Tg of P per year	Global: 11 Tg P/year; regional: 6.2 Tg P/year	Global: 100 Tg P/ year; regional: 11.2 Tg P/year	22.6; 17.5
	Nitrogen global: industrial and intentional fixation of N (Tg N/year)	0 Tg of N per year	62 Tg of N per year	82 Tg of N per year	190 Tg of N/ year
Land system change	Forested land as %age of original forest cover	100%	75%	54%	60%
Freshwater change	Human-induced disturbance of blue water flow	10.2%	9.4%	50%	18.2%

(continued)

Table 1.1 (continued)

Earth system process	Description	Preindustrial Holocene base value	Planetary boundary	Upper end of zone of increasing risk	Status
	Human-induced disturbance of (green) water available to plants (% land area deviating from preindustrial variability)	11.1%	9.4% (median preind.)	50%	15.8%
Atmospheric aerosol loading	Interhemispheric difference in Atmospheric Optical Depth (AOD)	0.1	0.03	0.25	0.076
Novel entities	Percentage of synthetic chemicals released to the environment without adequate safety testing	0	0	N/A	Transgressed

maintaining thriving ecosystems for water, land, and air, and limiting the global temperature rise to within 2° [42].

A circular economy on a global scale could keep conditions within planetary boundaries—an opportunity for technical sustainability management.

1.2 Social Dimension

There is no definitive and universal definition or conceptualization of social sustainability [43]. However, most of the 17 United Nations Sustainable Development Goals (SDGs) are linked to social aspects (see Fig. 1.1). This specifically stresses the importance of the intragenerational aspect of sustainability. Similar to the environmental dimension, progress has been made on a global scale for some of these social SDGs like improving the health situation. Here, 146 of 200 countries have already met or are on track to meet the mortality target for under 5-year-olds or the elimination of some tropical diseases. However, 8.4% of world population are still living in poverty (below $2.15/day), 9.2% suffer from chronic hunger and only one in six countries will achieve universal secondary school completion targets [4]. At current improvement rates, it will take around 230 years to end poverty while the first trillionaire is expected in the early 2030s [44] leading to huge intragenerational injustice. Health crises like COVID-19, environmental crises like climate change, geopolitical and warlike conflicts, inflation, and recessions are exacerbating deep-rooted inequities, with intense social repercussions ranging from polarization and declining levels of trust to social unrest. Social sustainability of societies has four dimensions [45]:

- Social cohesion—a sense of shared purpose, trust, and willingness to cooperate among members of a given group.
- Inclusion—allowing all people to thrive.
- Resilience—everyone, including the poor and marginalized, is safe and can withstand shocks without significant losses to the well-being of current and future generations.
- Process legitimacy—ensuring that policies and programs are designed and implemented so that they are accepted as fair and credible by all affected stakeholders.

These four dimensions are in line with human rights [46] and the human rights-based principles of participation, accountability, non-discrimination and equality, empowerment, and legality (aka PANEL) [47].

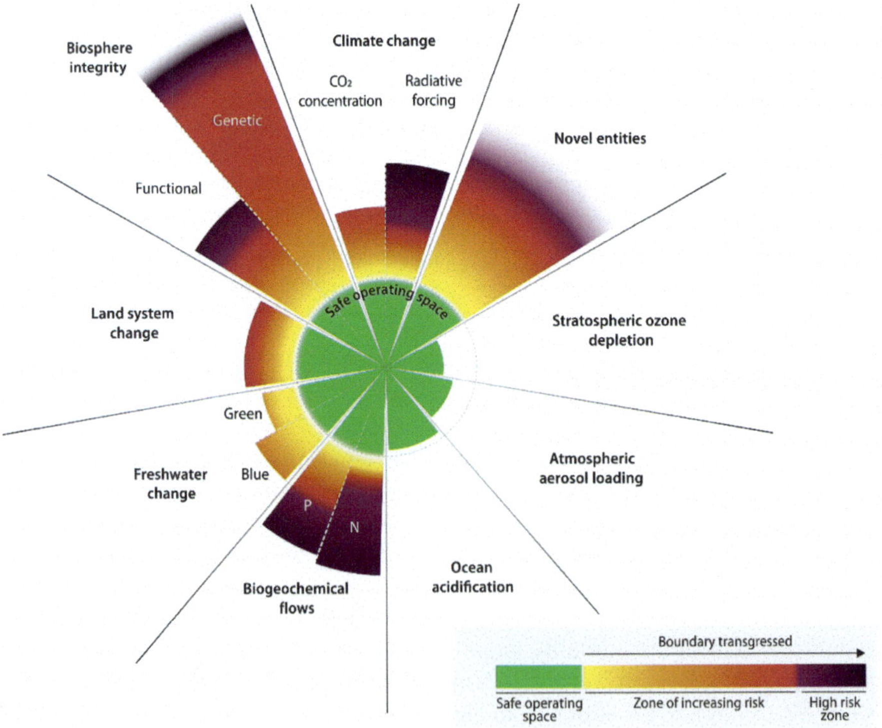

Fig. 1.7 Status of the planetary boundaries [40]

Social sustainability is a subject of most UN Sustainable Development Goals. It especially stresses intragenerational justice and is strongly linked to human rights.

The OECD differentiates between "current well-being" and "resources for future well-being" to reflect the sustainability definition, covering current and future generations [2]. "Current well-being" covers both material conditions as well as quality of life. Material conditions include income and wealth distribution, jobs and earnings, as well as housing indicators. Quality of life looks at health status, work-life balance, education and skills, social connections, civic engagement and governance, environmental quality, personal security, and subjective well-being. Future well-being assesses whether the natural, economic, human, and social capital can sustain well-being over time [48].

In reality, various challenges within but also beyond OECD countries exist.

Forced labor or modern slavery is an issue affecting 27.6 million people [49]. If forced marriage is included, up to 50 million people are affected [50]. Forced labor has two indicators: a lack of consent and a threat of penalty. A threat of penalty can include threats like destroying passports, retention of identity documents, confinement, violence, and/or non-payment of wages. Lack of consent can occur if the consent is not voluntary or not informed, for example, because:

- The contract is not intelligible for the worker.
- Verbal promises leading to consent are not part of the working contract.
- Important information is withheld prior to the consent like full costs of living/housing for the worker, real working hours or working days, etc.
- Deceiving potential workers about the conditions of work.

Bringing people into such unfavorable working conditions is human trafficking. Debt-bonded labor is a typical form of modern slavery where people are pushed into contracts where they first have to pay back excessive fees for recruitment, contract agents, transportation to the new job, accommodation, deposits, and food that are then deducted from their salary. Modern slavery is also present in Europe. For example, the police in Wales and England registered 5144 modern slavery offences in the financial year ending in March 2019 [51]. However, the estimated number of forced labors in the UK is much higher at around 122,000 in 2021 [50]. The Global Slavery Index [50] actually does not only have estimates for modern slavery for almost every country but also detailed documented examples. The highest prevalence of modern slavery is assumed to be in North Korea, Eritrea, Mauritania, and Saudi Arabia followed by Türkiye.

According to UNICEF "nearly 1 in 10 children are subject to child labor worldwide", with half of them working in hazardous conditions [52]. Child labor negatively impacts their education, health, and development, corresponds with economic exploitation, and often leads to sexual abuse and slavery. The US Bureau of International Labor Affairs published a list of 159 goods from 78 countries linked to child and/or forced work in farming and harvesting, (hazelnuts, natural rubber, cotton, grapes, blueberries, tobacco, olives, sugarcane, coffee, and many more), mining and building industry (stone, bricks, coal, diamonds, etc.), other industries (electronics, textile, leather, pyrotechnics, etc.) as well as pornography [53].

Human trafficking leads to forced labor in all parts of the world. A wide range of goods, in particular, in food/agriculture, mining, and building industry, are linked to child labor.

1.1 billion people or 18% of the world population live in multidimensional poverty and half of them are children under 18 years of age. 84% of poor people live in rural areas, 47.8% of poor people live in Sub-Saharan Africa, and 34.9% in South Asia. 485 million people across 110 countries live in severe poverty [54].

The Sustainability Society Index [55] covers social aspects with ten factors under the header of human well-being. The first three reflect basic needs: the number of undernourished people, the number of people with sustainable access to an improved water source, and the number of people with sustainable access to improved sanitation. The following three factors reflect health aspects: the life expectancy at birth in a number of healthy life years, air pollution, and its effects on humans (PM2.5), and surface water quality. The last four factors reflect the personal and social development opportunities: education measured as enrollment for primary, secondary, and tertiary education, gender equality measured by the gender gap index, fair income distribution measured as the ratio of income of the richest 10% to the poorest 10% of people and good governance as an average of the six governance indicators of the World Bank. Over time, the social dimension of the Sustainability Society Index has a slight positive trend—mainly due to improvements in meeting basic needs partly compensated by worsening of equality indicators [56]. According to this methodology, the worst social sustainability can be found in Somalia, Congo, Guinea-Bissau, Niger, and the Central African Republic, so only African countries are in the worst five.

The Sustainability Society Index covers 10 factors contributing to human well-being—covering basic needs, health, equity in education, gender gap, and income.

1.3 Economic Dimension

Economic sustainability can be described from a macro-economic, societal, and global perspective or from a micro-economic and organization-specific perspective.

On a macro-economic level, it includes the elements of development and growth. The United Nations Sustainable Development Goals (SDGs) "Good health and well-being" (SDG 3), "Develop sustainable, resilient and inclusive infrastructure, industrialization and innovation" (SDG 9), "Reduced inequalities" (SDG 10), and "Responsible consumption and production" (SDG 12) reflect on these elements. The macro-economic level may include intragenerational as well as intergenerational aspects. The intragenerational economic sustainability aspects are typically covered in the social dimension (see Sect. 1.2). However, intragenerational aspects have also negative impacts on the environmental and economic dimensions. The reason is that the wealth of individuals correlates with their negative environmental impact. For example, the richest 1% emit as much CO_2 as the poorest two-thirds of humanity [44]. The economic risk is the increase of monopolies and unfair competition if capital is concentrated in the hands of a few. For example, the richest 1% of the global population owns 43% of all global financial assets and a third of the global top 50 public corporations are either led by a CEO who is a billionaire or have a billionaire as the principal shareholder [44]. The current situation is far away from a "Palma of 1". The Palma ratio is the ratio between the national income share of the top 10% of households and the bottom 40%. A Palma of 1 means that the richest 10% of the population earn the same as the poorest 40% [57].

For intergenerational aspects debts in particular are seen as a risk to sustainability, as the ability of future generations to meet their needs is limited by excessive liabilities of debts from previous generations. Global debts of governments, businesses, and households reached $307 trillion in the second quarter of 2023 according to the International Finance Institute, leading—in the view of the International Monetary Fund—to the risk that at least 100 countries have to reduce their spending on health, education, and social protection to pay off debts [58]. However, if debts are taken on only to invest in infrastructure, education, and sustainability efforts to preserve and improve conditions for future generations, they may not be in conflict with intergenerational justice. The International Monetary Fund suggests that debts are sustainable as long as the government or organization "is able to meet all its current and future payment obligations without exceptional financial assistance or going into default" [59]. But the intergenerational aspects of economic sustainability go much further. In essence, from an economic perspective, it is important how much net capital is handed down from current to future generations. If the ratio of this Economic Sustainability Index (ESI) is over 100%, current generations have increased the stock of capital for future generations and thus increased economic sustainability.

The Economic Sustainability Index (ESI) measures how much net capital is inherited by the next generation. If ESI >100%, assets are increasing and debts for this increase are justified.

The capital in question includes the real capital RC (infrastructure, assets like factories), human capital HC (skills), natural capital NC (environment plus resources), structural capital SC (patents, intellectual properties, societal rules, and institutions) as well as intergenerational net debts D per year [60].

$$\mathrm{ESI} = \frac{(\mathrm{RC} + \mathrm{HC} + \mathrm{NC} + \mathrm{SC} - \mathrm{D})_{\text{handed down}}}{(\mathrm{RC} + \mathrm{HC} + \mathrm{NC} + \mathrm{SC} - \mathrm{D})_{\text{inherited}}} \tag{1.1}$$

This looks straightforward in theory, but the practical calculation is very difficult, as it must consider depreciation, discounting, monetarization, and general data availability issues.

On a micro-economic level, economic sustainability is "the process of allocating and protecting scarce resources, while ensuring positive social and environmental outcomes" which allows organizations to stay in business in the long-term [61]. This is based on the understanding that financially high-performing organizations might be successful in the short term, but only stay in business in the long-run if they positively contribute to social and economic well-being. Therefore, the economic sustainability of organizations looks at its

- Financial performance (e.g. cash, EBIT, return-on-invest, book value).
- Management of intangible assets (e.g. brand reputation).
- Interaction with the wider economy.
- Interaction with and management of social and environmental aspects.
- Value creation (economic, social, environmental).
- Long-term changes and economic resilience.

The importance of long-term prospects for businesses is decisive in areas where major transformation is required to stay in business. Global efforts towards carbon neutrality are leading to the foreseeable end of the use of fossil energy. In consequence, sectors like the automotive industry, high-energy-consuming, or inherent CO_2 emitting material industries like steel (see Sect. 6.4), cement (see Sect. 6.7), and fossil energy companies like oil and coal industries need to change their business models towards more sustainable value creation (see Sect. 3.8).

Sustainable organizations are profitable long-term with positive social/ environmental impacts. Transformation is needed for industries without sustainable business models (e.g. fossil fuel).

1.4 Governance Dimension

Corruption is another challenge of a more sustainable world. Transparency International defines corruption as the "abuse of entrusted power for private gain" and stresses that corruption erodes trust, weakens democracy, hampers economic development and further exacerbates inequality, poverty, social divisions, and the environmental crisis [62].

Their Corruption Perception Index (CPI) ranges from 0 (highly corrupt) to 100 (very clean). The global average value for CPI has been stable over the last decade at 43, with Scandinavia, New Zealand, and Singapore typically having the best scores [63].

Good governance practices can control corruption and have a positive impact on sustainable development. This is not only important on a country level by attracting sustainable investors but also on the level of organizations. Therefore, environmental, social, and (corporate) governance criteria—or ESG—are frequently used in the sustainability debate. Given the long-term higher likelihood that companies with a positive environmental and social added-value AND good governance are good investments, there are many ESG investment funds globally. They exceeded a global total of $3.1 trillion by the end of June 2023 [64]. Typically, these funds outperform their peer funds due to the good governance of companies and sustainable business models—unless there are interim economic shocks where for example fossil and other energy war profiteers have interim wins like in 2022.

According to the World Bank, governance "consists of the traditions and institutions by which authority in an organization or country is exercised. This includes the process by which governments and decision-makers are selected, monitored and replaced; the capacity of the organization to effectively formulate and implement sound policies and practices; and the respect of citizens/employees and society for the bodies that govern economic and social interactions among them".

The World Bank has developed the Worldwide Governance Indicators (WGI) report, based on six broad dimensions of governance for over 215 countries and territories: "(I) Voice and Accountability; (II) Political Stability and Absence of Violence; (III) Government Effectiveness; (IV) Regulatory Quality; (V) Rule of Law; and (VI) Control of Corruption. The WGI are composite governance indicators based on over 30 underlying data sources. These data sources are rescaled and combined to create the six aggregate indicators using a statistical methodology known as an unobserved components model. A key feature of this methodology is that it generates margins of error for each governance estimate. These margins of error need to be considered when making comparisons between countries and over time". The Word Bank monitors and publishes the WGI results for all 215 countries. The ratings depend on the specific targets but are typically lowest for Afghanistan, Eritrea, and highest for Scandinavian countries [65].

The index is also used in country credit rating systems. It is—correctly or incorrectly—subject to criticism of its methodology and use of subjective opinions of institutions. It is accused of having no direct insights into all countries, in particular, emerging ones.

ESG combines good governance with environmental and social sustainability, leading to better returns of organizations plus better credit ratings for countries with superior governance.

1.5 Other Dimensions

Beyond the typical three dimensions or pillars of sustainability—environment, social, and economic—or the three aspects of ESG (Environment, Social, and Governance), other dimensions of sustainability are also discussed. These include cultural sustainability, institutional sustainability, political sustainability [43], and human sustainability.

The UNESCO Hanghzou Declaration [66] sees culture as a potential fundamental enabler of sustainability because any development is "shaped by culture and local context, which ultimately also determines its outcomes". The declaration calls for culture "to be included as the fourth fundamental principle… of sustainability. The cultural dimension should be systematically integrated in definitions of sustainable development and well-being…".

The Sustainable Development Goal (SDG) 4 "Ensuring inclusive and equitable quality education and promote lifelong learning opportunities for all" is a cultural SDG; however, it is typically covered by the social dimension. Another important cultural aspect is "Sustainable consumption" (part of SDG 12)—an example that shows the ambivalence of culture as either an enabler or blocker of sustainability. If the culture is connected to a lifestyle of intensive resource consumption, it undermines sustainable consumption patterns (see Sect. 7.1). In addition, cultural sustainability is "rooted with respect to diversity" [67]. The United Nations Decade on Biodiversity combines environmental and cultural diversity by stating that the "Sustainable Development Goals acknowledge the natural and cultural diversity of the world, and confirm that all cultures and civilizations can contribute to sustainable development" [68]. SDG 5 on "Gender equality" and SDG 10 on "Reduced inequalities" capture some of the cultural aspects of diversity and inclusion even though these SDGs are also typically allocated to the social dimension.

Institutional sustainability in the narrow sense is about organizations having the capacity to continue to perform their functions over the long term. Another perspective on institutional sustainability goes beyond preserving organizations. It is about institutions'/organizations' "capability to coordinate human interaction in order to achieve specific sustainability goals" [69]. In this approach, institutions are involved in policy-making or implementation of sustainability. This is important because institutions and organizations can be enablers or blockers for sustainable development. There is a link between institutional sustainability and good governance.

Human sustainability is about maintaining and improving human capital in society and organizations. People are seen as valuable in their own right. Therefore, human sustainability fosters investments in health and education systems, access to services, nutrition, knowledge, and skills. "Human sustainability focuses on the importance of anyone directly or indirectly involved in the making of products, or provision of services or broader stakeholders (the human capital of the organization)". [70] This translates into corporate social responsibility (CSR) practices in regard to establishing workforce skill sets and diversity, equal opportunity, workplace diversity, and work–life balance as workplace principles. Additionally, strong and clearly defined corporate ethical positions based on multiple stakeholder perspectives must be promoted internally and externally to pursue "human welfare, equitable and just social practices and the fulfilment of human potential of all" [71]. Again, most of these principles are covered under the social dimension of sustainability.

Cultural and human sustainability are central parts of social sustainability but go partly beyond. Institutional sustainability has a strong overlap with good governance.

1.6 Conclusions

Multiple challenges for environmental, social, and economic sustainability as well as governance exist. The EU Commission's Joint Research Center (JRC) has provided an overview of sustainability indices, all of which suggest substantial sustainability gaps. An early analysis of the trade-offs between environmental, social, and economic sustainability in 2012 suggested that there is a positive correlation between social and economic sustainability in several countries. However, the study suggested trade-offs towards environmental sustainability (exemption Switzerland) leading to the situation that societies only significantly invest in environmental sustainability once environmental damage has already become very serious [55]. This has led to a "business as usual" or "too little too late" approach in environmental sustainability actions and a situation where the damage to the environmental dimension turns against both social and economic well-being. Obviously, only an approach that tackles all aspects of sustainability really leads to long-term sustainability.

The more recent 2023 Sustainable Society Index (SSI) edition's data [72]—covering the year 2019—suggests that there is no real correlation between economic, social, and environmental sustainability due to R-square being very low (R^2 being the coefficient of determination that shows how well the regression model fits with 1 being the perfect fit). However, it confirms a weak positive correlation between economic and social but a weak negative correlation between environmental and the other sustainability dimensions when considering all of the 213 countries monitored.

This could also be an artifact of the incompleteness of data for individual countries or biased by other factors like geography, governance, and culture. For example, if only all EFTA countries plus the UK are taken into consideration, there is a (weak) positive correlation between all three dimensions of sustainability, including the environment. On the other hand, if Lichtenstein is excluded from this calculation (because their economic well-being data exist for only one of five indicators—organic farming—and only for three of seven environmental indicators), the positive correlation between environmental and economic sustainability is close to non-existent.

Different from the past, there is no evidence anymore for a contradiction between environmental sustainability on the one hand and social and economic sustainability on the other hand.

Instead of business as usual in decision-making, a "giant leap" will be needed in the areas of poverty, inequality, empowerment, food, and energy. Otherwise, "we will condemn future generations to a dangerously destabilized planet; within a few decades the climate system is likely to cross multiple tipping points and social tensions are likely to increase". In contrast, the "giant leap" scenario with decisive actions significantly reduces the risks but will still fall short to deliver on two areas—inequality and climate change. For climate change, the 1.5 °C limit is modeled to be reached in the 2030s [15] or even earlier. It is important to remember how close the year 2023 is already to this limit—with 86 days exceeding 1.5 °C as well as devastating climate-related events [26]. So more than a "giant leap" will be needed and this requires immediate action and compromise from all. Such decisive actions will help mitigate a self-reinforcing downward spiral where environmental shortfalls heavily damage the social and economic dimensions. For example, the current trend for biodiversity damage will cost an estimated 2.7$ Trillion (2.3% of global GDP and 10% in low-income countries) [73], the damage of a global increase in temperatures to well below 2 °C (Paris agreement) is estimated to lead to a loss of 4.2% of global GDP, 2 °C to a loss of 11% global GDP, and 3.2 °C to a loss of more than 18% of GDP, with highest negative GDP impacts in South East Asia [74]. Therefore, it is important to recognize that if the right approach is chosen, all dimensions of sustainability will support each other and humanity will profit from such sustainable development. The higher economic success of sustainability investments shows that businesses also profit from this approach: according to some assessments, the top 100 companies offering products and services needed for sustainability have outperformed the market over these last few years [75].

Massive and immediate efforts towards sustainability are needed—otherwise the planet, societies, and the global economy will be dangerously destabilized.

In conclusion, humanity is facing multiple fundamental crises that require decisive, consistent, and immediate action on all levels. The following chapters outline how action is possible and necessary in all sectors and roles—as employees, managers, executives, or investors in industry, agriculture, transport, administration, education, communication, as well as consumers, politicians, and regulators, but also as judges and in many more roles.

References

1. United Nations, SDG, 25 September 2015. [Online]. Available: https://documents-dds-ny.un.org/doc/UNDOC/GEN/N15/291/89/PDF/N1529189.pdf?OpenElement. Accessed 27 Feb 2023
2. United Nations, Our Common Future (The Brundtland Report): World Commission on Environment and Development, UN, 1987
3. Stockholm Resilience Centre, Contributions to Agenda 2030, Stockholm
4. United Nations, The Sustainable Development Goals Report 2023: Special Edition—Towards a Rescue Plan for People and Planet, United Nations, 2023
5. G. Whitesides, Learning from Success: Lessons in Science and Diplomacy from the Montreal. Sci. Dipl. **9**(2) (June 2020)
6. UNEP, Montreal Protocol, 27 February 2023. [Online]. Available: https://www.unep.org/ozonaction/who-we-are/about-montreal-protocol. Accessed 27 Feb 2023
7. NOAA, Global Monitoring Laboratory—Earth System Research Laboratories, 27 February 2023. [Online]. Available: https://gml.noaa.gov/hats/graphs/graphs.html
8. Scripps, Institution of Oceanography, Scripps CO_2 Program, 27 February 2023. [Online]. Available: https://scrippsco2.ucsd.edu/data/atmospheric_co2/primary_mlo_co2_record.html. Accessed 27 Feb 2023
9. UNFCCC, 27 February 2023. [Online]. Available: https://unfccc.int/timeline/
10. Sceptical Science, Mason, 5 June 2020. [Online]. Available: https://skepticalscience.com/history-climate-science.html. Accessed 27 Feb 2023
11. Climate Group, A Call to the European Parliament and EU Governments to Support the European Commission's Proposed Phase-Out of New Diesel and Petrol Cars by 2035, 16 June 2022. [Online]. Available: https://www.theclimategroup.org/call-european-parliament-and-eu-governments-support-european-commissions-proposed-phase-out-new. Accessed 27 Feb 2023
12. Stockholm Environment Institute, Climate Analytics, E3G, International Institute for Sustainable Development and United Nations Environment Programme, The Production Gap: Phasing Down or Phasing Up? Top Fossil Fuel Producers Plan Even More Extraction Despite Climate Promises, Stockholm Environment Institute, 2023
13. UN, Paris Agreement, 15 December 2015. [Online]. Available: https://unfccc.int/sites/default/files/english_paris_agreement.pdf
14. IPCC, Synthesis Report of the IPCC Sixth Assessment Report (AR6) Summary for Policymakers, Intergovernmental Panel on Climate Change, Geneva, 2023
15. J. Bernstein, D. Collste, S. Dixson-Declève and N. Spittler, SDGs for All: Strategic Scenarios, Earth4All System Dynamics Modelling of SDG Progress, 2023
16. Copernicus Climate Change Service (C3S), In 2024, the World Experienced the Warmest January on Record, 8 February 2024. [Online]. Available: https://climate.copernicus.eu/copernicus-2024-world-experienced-warmest-january-record
17. Copernicus Climate Change Service (C3S), Warmest February on Record, the 9th Consecutive Warmest Month, 2024
18. NOAA/NCEI, Climate at a Glance Global Time Series [Online]. Available: https://www.ncei.noaa.gov/access/monitoring/climate-at-a-glance/global/time-series. Accessed 30 Jan 2024

19. J. Song, G. Tong, J. Chao, J. Chung, M. Zhang, W. Lin, T. Zhang, P.M. Bentler, W. Zhu, Data Driven Pathway Analysis and Forecast of Global Warming and Sea Level Rise. Sci. Rep. **13**, 5536 (2023)
20. A.E. Raftery, E. Zimmer, D.M.W. Frierson, R. Startz, P. Liu, Less Than 2 °C warming by 2100 Unlikely. Nat. Clim. Chang. **7**, 637–641 (2017)
21. T.M. Lenton, L. Laybourn, D.I. Armstrong McKay, S. Loriani, J.F. Abrams, S.J. Lade, J.F. Donges, M. Milkoreit, S.R. Smith, E. Bailey, T. Powell, C.Z.C.B.J.B.J.D. Fesenfeld, *Global Tipping Points Report 2023* (University of Exeter, Exeter, 2023)
22. IPCC, International Panel on Climate Change, Global Warming of 1.5°C. An IPCC Special Report on the Impacts of Global Warming of 1.5°C Above Pre-Industrial Levels and Related Global Greenhouse Gas Emission Pathways, in the Context of Strengthening the Global Response to the Threat of Climate Change, 2018
23. IPCC Core Writing Team, H. Lee and J. Romero (eds.), Contribution of Working Groups I, II and III to the Sixth Assessment Report of the Intergovernmental Panel on Climate Change, in *Climate Change 2023: Synthesis Report* (IPCC, Geneva, 2023), pp. 35–115
24. Climate Council, Impacts at 1.5 and 2 Degrees of Warming [Online]. Available: https://www.climatecouncil.org.au/resources/impacts-degrees-warming/. Accessed 3 Dec 2023
25. SRU, German Advisory Council on the Environment, *Using the CO_2 Budget to Meet the Paris Climate Targets* (SRU, Berlin, 2020)
26. UNEP, *Broken Record—Temperatures Hit New Highs, Yet World Fails to Cut Emissions (Again)* (United Nations Environment Programme, Nairobi, 2023)
27. Mercator, Research Institute on Global Commons and Climate Change, Remaining Carbon Budget [Online]. Available: https://www.mcc-berlin.net/en/research/co2-budget.html. Accessed 2 Feb 2024
28. M. Exposito-Alonso, T.R. Booker, L. Czech, T. Fukami, L. Gillespie, S. Hateley, C.C. Kyriazis, P.L.M. Lang, L. Leventhal, D. Nogues-Bravo, V. Pagowski, M. Ruffley, J.P. Spence, S.E. Toro, C.L. Arana, Z.E. Weiß, Genetic Diversity Loss in the Anthropocene. Science **377**, 1431–1435 (2022)
29. World Economic Forum, Nature Risk Rising: Why the Crisis Engulfing Nature Matters for Business and the Economy, Geneva, 2020
30. UNEP, Convention on Biological Diversity, Secretariat of the Convention on Biological Diversity—United Nations Environment Programme, Montreal, 1992
31. R. Dirzo, G. Ceballos, P.R. Ehrlich, Circling the Drain: The Extinction Crisis and the Future of Humanity. Philos. Trans. **377**(1857) (15 August 2022)
32. A.L.S. Meyer, J. Bentley, R.C. Odoulami, A.L. Pigot, C.H. Trisos, Risks to Biodiversity from Temperature Overshoot Pathways. Philos. Trans. **377**(20210394), (2022)
33. United Nations, Biodiversity—Our Strongest Natural Defense Against Climate Change [Online]. Available: https://www.un.org/en/climatechange/science/climate-issues/biodiversity. Accessed 2 Dec 2023
34. BBNJ, "Draft agreement under the United Nations Convention on the Law of the Sea on the conservation and sustainable use of marine biological diversity of areas beyond national jurisdiction," United Nations, New York, 2023
35. Convention on Biological Diversity, Final Text of Kunming-Montreal Global Biodiversity Framework, Montreal, 2022
36. R.A.M. Peer, C.M. Chini, Historical Values of Water and Carbon Intensity of Global Electricity Production. Environ. Res. Infrastruct. Sustain. **1**(025001), (16 July 2021)
37. S. Erhart, K. Erhart, *Environmental Ranking of European Industrial Facilities by Toxicity and Global Warming Potentials* (NaturePortfolio, Ispra, Italy, 2023)
38. J. Garnett, C. Halsall, A. Vader, H. Joerss, R. Ebinghaus, A. Leeson, P.M. Wynn, High Concentrations of Perfluoroalkyl Acids in Arctic Seawater Driven by Early Thawing Sea Ice. Environ. Sci. Technol. **56**, 11049–11059 (26 July 2022)
39. World Health Organization, Targets of Sustainable Development Goal 3 [Online]. Available: https://www.who.int/europe/about-us/our-work/sustainable-development-goals/targets-of-sustainable-development-goal-3

40. K. Richardson, W. Steffen, W. Lucht, J. Bendtsen, S.E. Cornell, J.F. Donges, M. Drüke, I. Fetzer, G. Bala, W. Von Bloh, G. Feulner, Earth Beyond Six of Nine Planetary Boundaries. Sci. Adv. **9**(37) (13 September 2023)
41. D.I. Armstrong McKay, A. Staal, J.F. Abrams, R. Winkelmann, B. Sakschewski, S. Loriani, I. Fetzer, S.E. Cornell, J. Rockström, T.M. Lenton, Exceeding 1.5°C Global Warming Could Triggermultiple Climate Tipping Points. Science **377**, eabn7950 (9 September 2022)
42. Circle Economy, *The Circularity Gap Report 2023* (Circle Economy, Amsterdam, 2023)
43. J. McGuinn, E. Fries-Tersch, M. Jones, C. Crepaldi, M. Masso, I. Kardarik, M. Samek Lodovici, S. Drufuka, M. Gancheva and B. Geny, *Social Sustainability—Concepts and Benchmarks. Study on Request of European Parliament's Committee on Employment and Social* (European Parliament, Luxembourg, 2020)
44. Oxfam International (Lead authors: Rebecca Riddell, Nabil Ahmed, Alex), Inequality Inc.—How Corporate Power Divides Our World and the Need for a New Era of Public Action, Oxford, 2024
45. P. Barron, L. Cord, J. Cuesta, S.A. Espinoza, G. Larson, M. Woolcock, *Social Sustainability in Development—Meeting the Challenges of the 21th Centuries* (International Bank for Reconstruction and Development/The World Bank, Washington, 2023)
46. United Nations Universal Declaration of Human Rights, Paris, 1948
47. European Network of National Human Rights Institutions, Human Rights-Based Approach [Online]. Available: https://ennhri.org/about-nhris/human-rights-based-approach/#:~:text=Principles,and%20Equality%2C%20Empowerment%20and%20Legality. Accessed 3 Dec 2023
48. OECD, Measuring Distance to the SDG Targets 2019: An Assessment of Where OECD Countries Stand, Paris, 2019
49. International Labor Organization, IOM, Global Estimates of Modern Slavery: Forced Labour and Forced Marriage, Geneva, 2022
50. Walk Free, *The Global Slavery Index* (Minderoo Foundation, 2023)
51. Office for National Statistics, Modern Slavery in the UK, London, March 2020
52. UNICEF, Child Labour [Online]. Available: https://www.unicef.org/protection/child-labour. Accessed 5 Mar 2023
53. US Department of Labor, List of Goods Produced by Child Labor or Forced Labor. Bureau of International Labor Affairs [Online]. Available: https://www.dol.gov/agencies/ilab/reports/child-labor/list-of-goods-print. Accessed 25 Oct 2023
54. United Nations Development Program (UNDP), Oxford Poverty and Human Development Initiative (OPHI), Global Multidimensional Poverty Index 2023—Unstacking Global Poverty: Data for High Impact Action, 2023
55. M. Saisana, D. Philippas, *Sustainable Society Index (SSI): Taking Societies' Pulse Along Social, Environmental and Economic Issues* (EU Commission—Joint Research Center, Ispra, Italy, 2012)
56. T.H. Köln, Sustainable Society Index, Summary Report 2018, Köln, 2018
57. B.L. Kron, Economic Inequality, 2024
58. World Economic Forum, What Is 'Global Debt'—And How High Is It Now? 2 October 2023 [Online]. Available: https://www.weforum.org/agenda/2023/10/what-is-global-debt-why-high/. Accessed 30 Oct 2023
59. D. Hakura, What Is Debt Sustainability? Fin. Dev., 60–61 (30 September 2020)
60. P. Ederer, P. Schuller, S. Willms, The Economic Sustainability Indicator, in *Handbook of Intergenerational Justice* (Edward Elgar Publishing, Cheltenham, 2006), pp. 129–147
61. D. Doane and A. MacGillivray, Economic Sustainability—The Business of Staying in Business, 2001
62. Transparency International, What Is Corruption? [Online]. Available: https://www.transparency.org/en/what-is-corruption. Accessed 1 Nov 2023
63. Transparency International, Corruption Perceptions Index [Online]. Available: https://www.transparency.org/en/cpi/2022. Accessed 1 Nov 2023
64. Morgan Stanley, Sustainable Reality—Sustainable Funds Return to Outperformance in First Half of 2023, Morgan Stanley, 2023

65. World Bank, Worldwide Governance Indicators, 29 September 2023 [Online]. Available: https://datacatalog.worldbank.org/search/dataset/0038026/Worldwide-Governance-Indicators. Accessed 24 Oct 2023
66. UNESCO United Nations Educational, Scientific and Cultural Organization, The Hangzhou Declaration—Placing Culture at the Heart of Sustainable Development Policies, Hangzhou, China, 2013
67. T. Meireis and G. Rippl, *Cultural Sustainability* (Routledge, New York, Oxon, 2019)
68. United Nations Decade on Biodiversity—2011–2020—Sustainable Development Goals [Online]. Available: https://www.cbd.int/2011-2020/about/sdgs. Accessed 1 Nov 2023
69. S. Pfahl, Institutional Sustainability. Int J Sustain Dev **8**(1–2), 80–96 (2005)
70. RMIT, The Four Pillars of Sustainability, 2017 [Online]. Available: https://www.futurelearn.com/info/courses/sustainable-business/0/steps/78337
71. S. Benn, D. Dunphy, M. Griffiths, *Organizational Change for Corporate Sustainability* (Routledge, London, 2014)
72. T.H. Köln, Sustainable Society Index, 2023 Edition, ssi-latest.xlsx, Downloaded October 25 2023 from https://ssi.wi.th-koeln.de [Online]. Available: https://ssi.wi.th-koeln.de/index.html
73. J.A. Johnson, G. Ruta, U. Baldos, R. Cervigni, S. Chonabayashi, E. Corong, O. Gavryliuk, J. Gerber, T. Hertel, C. Nootenboom, S. Polasky, *The Economic Case for Nature—A Global Earth-Economy Model to Assess Development Policy Pathways* (The World Bank, Washington, 2021)
74. J. Guo, D. Kubli, P. Saner, *The Economics of Climate Change: No Action Not an Option* (Swiss Re Institute, Zurich, 2021)
75. M. Scott, 100 Most Sustainable Companies of 2023 Still Flourishing in Tumultuous Times, Corporate Knights—Sustainable Economy Intelligence, 18 January 2023 [Online]. Available: https://www.corporateknights.com/rankings/global-100-rankings/2023-global-100-rankings/2023-global-100-most-sustainable-companies/#more-story. Accessed 3 Nov 2023

Chapter 2
Fundamentals: Holistic Life Cycle Thinking

Abstract It is crucial to focus efforts towards sustainability on the correct targets. Most sustainability issues and mitigating actions are linked to products and services offered by organizations/countries or demanded by customers/societies. A good starting point is to identify areas of biggest concern as well as assessing the benefit of interventions based on their sustainability. Potential issues and opportunities along the product's life cycle from environmental, social, and economic perspectives should be determined. The chapter describes the necessary tools to identify areas of particular relevance, focusing on Life Cycle Assessment, Social Life Cycle Assessment, and Life Cycle Costing.

Keywords Life Cycle Assessment · Social-LCA · Life Cycle Costing · Total Cost of Ownership · Life Cycle Management · Allocation · Discounting · Hot spot · Impact category · Data quality

It is crucial to focus the necessary massive and immediate efforts toward sustainability on the correct aspects. Most sustainability issues and mitigation actions are linked to products and services offered by organizations/countries or demanded by customers/societies. A good place to start identifying the hot spots of biggest concern as well as assessing alternative actions based on their sustainability is to assess the potential issues and opportunities along the product's life cycle—from environmental, social, and economic perspectives.

2.1 Life Cycle Thinking

Life cycle thinking is the fundamental approach in this context. "Life cycle thinking is the way of thinking of the consequences in the environmental, economic and social effects of a product throughout its entire life" [1, 2].

A product life cycle covers several processes. It starts with the extraction of raw materials and generation of energies that flow into the required parts and final product

© The Author(s), under exclusive license to Springer Nature Switzerland AG 2024

W. Schmidt, *Solutions For Sustainability Challenges*,
https://doi.org/10.1007/978-3-031-63624-0_2

manufacturing—including all process and logistic steps. This upstream part of the life cycle is sometimes called "cradle-to-gate" i.e. every step until the finished product leaves the final manufacturing gate.

The use or consumption of the product can be very complex, and life cycles may also include maintenance steps. For mobility products like vehicles and trains, a study that only looks at the direct emissions during the use of the mobility carrier moving, is called "tank-to-wheel". If in addition the emissions of energy generation are covered, the study is called "well-to-wheel".

Finally, the end-of-life (EOL) phase includes all collection, waste processing, recycling, recovery, and disposal processes. The whole life cycle from raw material extraction to final disposal is called "cradle-to-grave". In case of a circular approach where the materials flow back, a life cycle is called "cradle-to-cradle". Transportation processes allow materials and products to flow along this life cycle between individual processes (see Fig. 2.1).

The importance of each Life Cycle phases varies between different products and services depending on where the most sever impacts ("hot spots") occur. There are several general rules that apply to typical products and services unless dedicated actions are carried out to mitigate their effects.

> Holistic Life Cycle Thinking needs to look at all processes and their input/outputs including raw material extraction to production, use, end-of-life operations, energy and transportation.

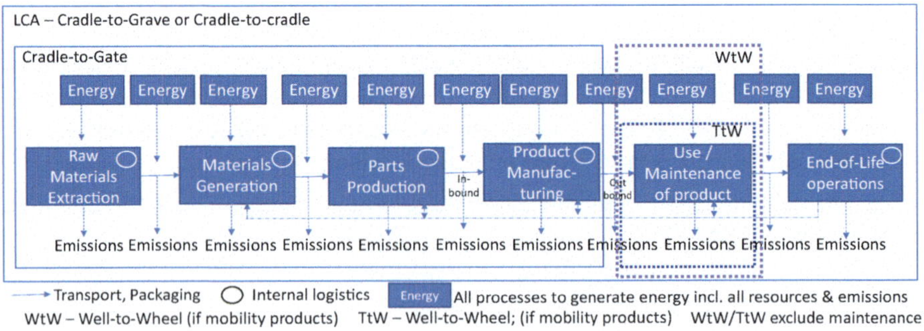

Fig. 2.1 Life cycle of a product with different system boundaries. ©*Wulf-Peter Schmidt 2024. All Rights Reserved*

2.1.1 Production Sensitive Products and Services

Production sensitive products and services are linked in particular to materials that require a lot of energy/water/chemicals to generate (like Gold, Titanium, concrete, rare earth minerals) and/or where the processing of these materials corresponds with the use of lots of energy/ water/chemicals/emissions (like wafer or battery production). Examples include jewels, windmills, and solar panels. Here, the other phases in the life cycle have little to no significance or even a positive impact to sustainability. For other production intensive products like electronics, buildings, meat or dairy production, and transport vehicles, the relative importance of the production phase depends on how efficient and how close to zero emissions their usage is. For example, the relative importance of the production phase for an electric vehicle is relatively high if the electric vehicle is charged predominantly with renewable electricity and relatively low for conventional vehicles fueled with fossil energy. The material losses from material production to the final product must also be considered.

On a global scale, the highest absolute carbon emissions are linked to the production of steel (3.3 Gt CO_2 or 31% of all material production-related carbon emissions), followed by cement (2.6 Gt) [3]. In relative terms, the carbon-intensity (kg carbon emissions per kg material) as well as other environmental emissions are available in the professional LCA software packages as well as in public databases. It is remarkable that there is a correlation between cost and carbon-intensity for several materials; for example, gold, titanium and rare earth minerals are linked to significantly higher emissions and costs per kg of material than paper or polypropylene.

The carbon-intensity of material production depends on the technologies chosen, the ease of access to resources (e.g. concentration of mineral in ore as well as how deep ores occur), the share of recycled content, and the type of energy used in mining and material production. The carbon-intensity varies widely between countries of origin: Aluminum produced in Norway has a much lower carbon-intensity than in China or South Africa mainly due to the use of renewable versus non-renewable energy sources [4]. Responsible and sustainable decisions in regards to sourcing should take this into consideration where primary data is available—at least for major materials and investments.

> Typically, the phase from raw material extraction to final production is most important for high-energy/high-cost products (e.g. Gold, electronics) or for special impacts (e.g. meat).

Social issues are often linked to the raw material extraction phase especially in developing economies. 70% of child labor occurs in agriculture (food, natural rubber) while also being prevalent in mining in very poor countries. In manufacturing, child labor is present for example in the textile sector.

The environmental impacts of production sensitive products can be addressed by sufficiency (see Sect. 7.1), substitution of high-impact materials, more efficient or alternative processes and designs, a circular economy, and other means. The social impacts depend on individual sourcing decisions as well as political measures.

2.1.2 Usage Sensitive Products and Service

Usage sensitive products are those products that consume larger amounts of energy or chemicals during their usage than during their production or other life cycle phases. Long-living products like conventional houses, traditional mobility products as well as longer-living electronics are key examples. The relatively high share of the use phase may collapse with new technologies like net zero, passive houses, or electric vehicles Fig. 2.2.

For conventional vehicles, around 80% of the energy demand and carbon footprint is linked to the use of the vehicle, a figure that rises even more for heavy duty vehicles, while the supply chain is close to 20% [5, 6]. Vehicle manufacturing and end-of-life operations are essentially negligible in a vehicle's complete life cycle assessment (LCA) and less than the overall uncertainty of the LCA. For Battery Electric Vehicles (BEV), supply-side emissions and usage phase emission are both around 50%, if taking into account the average carbon footprint of electricity in Europe at the beginning of the 2020s. The expectation is that BEVs will be production (supply chain) sensitive products by the mid of the century—a situation that is already the case in Scandinavia due to the dominating position of renewable electricity in its grid mix (Fig. 2.2).

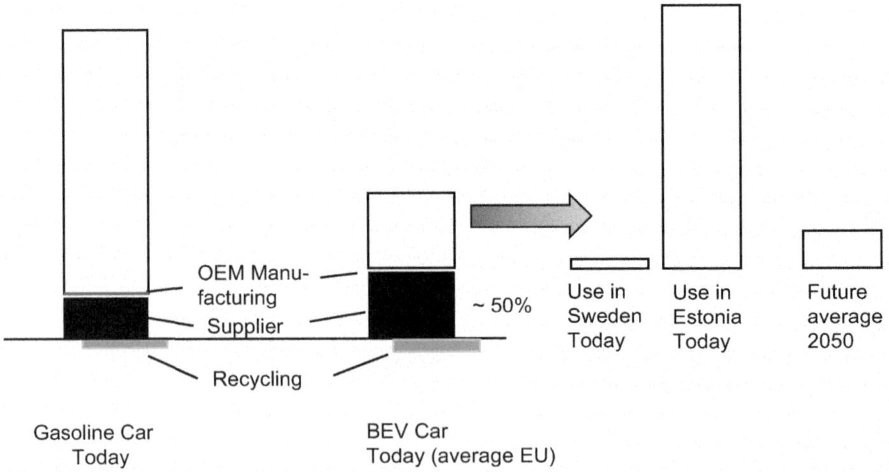

Fig. 2.2 Life cycle carbon footprint of gasoline and battery electric vehicles; based on [5]. ©*Wulf-Peter Schmidt 2024. All Rights Reserved*

Typically, the use phase of products is most relevant for repeatedly, long-living used products and services with a high consumption of non-renewable energy, water or chemicals.

The situation is similar from a social sustainability perspective. Conventional vehicles have a high oil demand in their use phase (refined to gasoline or diesel) which is linked to serious issues. This is especially true if the oil is sourced from countries with an oil-based economy where oil accounts for 60–95% of their exports. These countries typically have significantly higher shares of poverty and corruption but also child mortality and labor poor health care. They are typically authoritarian or failed states [7]. Similar issues can be observed for mineral-based economies. This explains why BEVs must be carefully audited from the initial suppliers of battery minerals like cobalt or nickel. Other typical use phase sensitive areas for child labor are domestic, food, and refreshment services.

2.1.3 End-of-Life Sensitive Products and Services

Only a few end-of-life sensitive products and services are known. These are typically those where long-term disposal concerns exist (e.g. nuclear products), where the end-of-life processes are damaging (e.g. hazardous chemical waste), or where no end-of-life solution exists (nuclear products in most countries).

However, many more products are sensitive to a circular economy. While the end-of-life phase itself typically has small negative impacts, major positive impacts for production-sensitive products could be achieved if virgin materials are replaced by recycled materials—as long as the environmental and social impacts of recycling processes are relatively low. These impacts of recycling depend on several factors such as impurities and effort in recycling, the quality of recycled products, what is replaced and the working conditions.

2.1.4 Transportation Sensitive Products and Services

Products are especially transportation sensitive if aviation or climatization is required during the supply chain (e.g. flying frozen food) of a product or as part of the service (e.g. tourism). For transportation sensitive products and services, the degree to which sustainability limits are exceeded increases with the distance of transportation. This differs from most products that are typically not transport sensitive i.e. where the impact of distances in supply chains is negligible in the overall environmental impacts over the life cycle. This is especially the case if there are relatively high emissions in production or use (e.g. electronics, vehicles, or food). This typically relatively low impact of transportation is driven by the high number of products transported per transport unit (e.g. number of products per container and number of containers per ship). For example, the difference between beef sourced from a farmer next door and beef from far away is less than 1% because the production of beef is so polluting [8].

The End-of-life phase and transportation are often less relevant unless there are specific large emissions (aviation, hazardous waste) or large benefits (e.g. replacing virgin material).

Again, sufficiency is the best approach to address the sustainability of transportation sensitive products and services (see Sect. 7.1), followed by alternative transportation modes for example by a shift of transportation from air freight to trains. In practice, these shifts are limited by capacity, required speed, reliability, and flexibility. Technological solutions like so-called "Sustainable Aviation Fuels" or Green-hydrogen fueled hydrofoil cargo ships are costly, limited in availability, still evolving and their sustainability is contested (see for example [9, 10]).

2.1.5 Tools Supporting Life Cycle Thinking

As general rules for hot spots might be too broad and there are too many specific conditions that may alter these conclusions, it is better to look closely into the details with one of the available tools.

Tools for assessments include Life Cycle Assessment (LCA) for environmental aspects [11, 12], Life Cycle Costing for economic aspects (LCC) [13, 14] and Social Life Cycle Assessment (S-LCA) for social aspects [15]. Methods like the Life Cycle Sustainability Assessment, Product Sustainability Assessment [16] or the Product Sustainability Index [17] follow the same principles by combining all dimensions of sustainability. More simplified approaches like checklists or the Environmental Multi-Criteria Requirement Matrix (MCRM) can be considered life cycle thinking as long as they sufficiently cover important aspects of the life cycle as outlined in Fig. 2.1.

The system boundaries of a full Life Cycle Assessment [12], or "cradle-to-grave" approach, typically cover all processes from raw material extraction through to product use in addition to the end-of-life operations like reuse, recycling, and recovery, including the flow of the resulting end-of-life products back into the same or other product life cycles. Emissions from final disposal are also covered. The ISO 14040/44 series has internationally harmonized the method since 1997. It also forms the basis of the narrower Product Carbon Footprint according to ISO 14067 that focuses only on greenhouse gas emissions (and often only to cradle-gate) and regulatory developments in the context of the Product Environmental Footprint in the European Union [18].

Based on the general LCA methodology, "cradle-to-cradle" is a trademarked term for an approach that is described as a biomimetic approach where products/materials at their end-of-life flow back as "nutrients" into either technical or biological systems [19].

Life Cycle Costing (LCC) is "an economic method of project evaluation in which all costs arising from owning, operating, maintaining, and ultimately disposing of a project are considered to be potentially important to that decision" [14]. The concept of an LCC originated in 1933 from the United States of America General Accounting Office's (GAO) request for tractor's costs from tenders from a life cycle perspective. Today, LCC studies may be complementary to an LCA, assessing "all costs associated with the life cycle of a product that is directly covered by any one or more of the actors in the product life cycle (e.g., supplier, manufacturer, user or consumer, or End of Life actor) with complementary inclusion of externalities that are anticipated to be internalized in the decision-relevant future" [20].

Social Life Cycle Assessment considers the "positive and negative impacts of the product life cycle" for workers, consumers, local communities, society, and value chain actors like suppliers in terms of health and safety, human rights, working conditions, socio-economic repercussions, cultural heritage and governance and/or other aspects covered under human well-being [15]. It is the youngest and least mature life cycle method [21].

> Product Life Cycle Thinking methods like LCA, LCC, S-LCA, LCSA cover various impacts on sustainability from raw material extraction to production, use and end-of-life operations.

"Life Cycle Management (LCM) is the organizational dimension of the life cycle approach. It is an integrated concept to change the total life cycle of goods and services toward a more sustainable production and consumption. LCM uses various procedural and analytical tools for different applications and integrates economic, social, and environmental aspects into an institutional context" [15, 22].

The combination of LCAs, LCCs, and S-LCAs in a Life Cycle Sustainability Assessment has been proposed and promoted by scientific communities in the UNEP/SETAC Life Cycle Initiative [23].

2.2 Life Cycle Assessment (LCA)

Product and Service-oriented Life Cycle Assessments (LCAs) can be carried out by various methods. Normally, "full LCAs" follow ISO 14040/14044:2006 and cover the full life cycle (see Fig. 2.1). There are simplified, screening, or streamlined LCAs that have a reduced scope (e.g. only cradle-to-gate meaning from raw material extraction to final product creation) or cover a limited number of environmental dimensions (e.g. only the carbon footprint, so exclusively looking at those emissions relevant for climate change). Also, the types of data can be different. Industry average data, or specific, primary data referring to the specific supply chain can be used. Similarly, there are Product Environmental Footprints (PEF) that follow specific rules for the product category agreed by the European Union [24] or LCA rules for specific products on an international level by ISO (e.g. ISO 20915:2018 for steel products). ISO 14025 applies to LCAs in so-called Environmental Product Declarations for external communication. The following chapters apply to all of these types of environmental LCA methods and applications in principle.

2.2.1 Goal and Scope Definition

2.2.1.1 Goal, System Boundaries, Data Requirements, and Functional Unit

Before starting any study, the goal of a study needs to be clarified as this determines methodological choices. Goals can include for example:

- The identification of hot spots (areas of highest environmental impacts).
- Decision support for material or supplier selection (determines whether industry average LCA data is sufficient of specific supplier data is needed).
- Decision support for strategic decisions for future technologies (may require a consequential LCA approach as described below).
- A comparative assertion between products disclosed or not disclosed to the public (an external third-party validation of the LCA is required if such results are published according to ISO14040:2006).

As stated in ISO 14044:2006 "The scope shall clearly specify the functions (performance characteristics) of the system being studied."

> The goal and scope definition of an LCA determines methodological choices such as the required type of data, the appropriate system boundaries, the requirements for validation.

System boundaries have to be introduced due to the complexity of life cycles and practical limits. However, the applied cut-off criteria have to be consistent and prevent relevant parts not being reflected.

The definition of the system with its boundaries typically looks at all processes connected to the goal of the study, from the extraction of raw materials to the final end-of-life operations (Fig. 2.1). This "product system" is composed of individual "unit processes" that represent the smallest element or activity for which data can be collected. The product system includes "foreground processes" (close to the studied product/service for which specific data is often collected) and "background processes" (further upstream or downstream processes for which generic data from databases is often applied). Capital goods—i.e. construction of buildings or infrastructure—are only included if they are the subject/goal of the study because their contribution is typically far from significant unless they have a relatively short lifetime and low utilization rate. LCA studies often also include cut-off thresholds below which the manufacturing of minor pre-product materials is excluded (e.g. based on a 1% threshold), however, this needs to be carefully checked for potential special environmental hot spots, justified, and explained. Equally, LCA studies often exclude processes that are not (solely) attributional to the product system– like emissions of consumer shopping or employee commuting (although employee commuting might be part of organizational Life Cycle studies) (see Sect. 2.6).

In goal and scope definitions, requirements for the type and quality of data are a necessary and helpful step. For example, the use of secondary, global industry average data could be considered more representative and pertinent for a study comparing general technology. Primary data covering the specific local or regional suppliers might make sense if specific products of different suppliers are compared—at least for some key aspects (see for example a four-level proposal discussed for automotive LCAs on UN-ECE level [25]) and if audited. Also, minimum data quality goals or objectives need to be defined and ideally translated into data quality indicators. Key is that the data quality is appropriate for the goal of the study. Depending on the goal and scope, the relevant impact categories must be selected with justification.

The "functional unit"—especially for comparative LCAs—defines the functions of the system and therefore also the object(s) of the study to enable a quantifiable comparison between the different products, processes, or service options. It should reflect their core characteristics such as functionality, technical quality, time aspects, additional services, and potentially also customer perceptions in terms of image, esthetics, and affordability. Instead of already determining a specific product (e.g. a car, a shirt, an oven), it should reflect the function provided by the compared life cycle systems (e.g. moving a certain number of people, covering a body of a certain size for a specified time and for a specific purpose like sports or festive events, keeping a defined room at a certain temperature). When it comes to LCAs the term "product" typically includes "services" as a product option.

> The "functional unit" enables a quantifiable comparison between different options and needs to define core characteristics of the required functions of all compared systems.

The "reference flow" translates this functional unit into the constituent product flows that make up to the functional unit. For example, if the objective of covering a body for a specific time can be achieved by two shirts from one supplier as opposed to three shirts from another supplier, the reference flows will be respectively two and three shirts.

2.2.1.2 Attributional, Consequential, and Simplified LCAs

In an "attributional LCA", these system-wide flows are attributed to the delivery of a specified amount of the functional unit. An attributional LCA models the system in a linear fashion where the flows all increase linearly with the functional unit. The absolute magnitude of the functional unit or how often the functional unit might be needed is of limited importance (e.g. emissions per g, kg, or ton). A "consequential LCA", in contrast, is an estimate of the system-wide change in pollution and resource flows that will result from a change in the level of the functional units produced. It might even be the case that the results depend on the magnitude of the change respective to the magnitude of the level of production. For example, if the goal of an LCA is to estimate the environmental impact of a major technology change in a country or region, this may have substantial consequences for the structure of the electricity system. Since the consequences do not scale linearly with the magnitude of the change, the results of a consequential LCA are easier to interpret if the functional unit reflects the magnitude of the change investigated [26].

An "attributional LCA" includes all processes that are expected to contribute significantly to the studied functional unit (e.g. average electricity sources). In a "consequential LCA", the system boundaries need to include the potentially altered future state (e.g. new electricity sources or electricity sources during times of high demand). The definition of system boundaries depends on how the markets can be expected to react to the studied change, taking into account market constrains, impact to other systems, etc. This leads to a situation where marginal processes are typically included—like new energy plants needed to satisfy the additional demand assumed in the "consequential LCA". More precisely, "the marginal processes are likely to be the unconstrained processes with the lowest expected marginal production costs. An exception to this rule is when the total market for that good is shrinking faster than the replacement rate of production capital in that market. In this case, the marginal process is likely to be the existing process with the highest expected marginal production costs" [26].

"Attributional" LCAs have linear flows based on current, average processes. "Consequential" LCAs reflect the magnitude of an investigated change by using marginal or future processes.

The goal and scope definition is also the place where potential simplified approaches are determined. Simplifying LCAs is a highly disputed approach that requires careful justification, implementation, and reflection of the use/interpretation of the simplified LCA results. A simplified LCA is an "application of the LCA methodology for a comprehensive screening assessment i.e. covering the whole life cycle but in a superficial way e.g. using generic data (qualitative and/or quantitative), standard modules for transportation or energy production, followed by a simplified assessment i.e. focusing on the most important environmental aspects … and/or stages of the life cycle and/or phases of the LCA and a thorough assessment of the reliability of the results" [27]. In other words, the idea is to identify hotpots before starting a detailed LCA. Hotspot identification is based on a qualitative/semi-quantitative/quantitative screening of the whole life cycle and all inputs/outputs, trying to identify the most important "elementary flows" and life cycle stages with the intention to focus on those. Expected data gaps are also identified. With such a pragmatic approach, the resource needs to conduct an LCA are reduced but the reduced scope will limit the reliability and usefulness of the LCA results.

The screening and simplification process can be organized in an iterative way. The screening indicators may change during the iteration steps first, qualitative or semi-quantitative assessments (ABC-rating for environmental severity and XYZ-rating for different levels of quantities) are conducted based on engineering judgment, expert panels, or previous studies to identify important areas. In a following iteration step, the identified initial, broader hotspot areas are further narrowed down using quantitative screening indicators that may indirectly influence several environmental dimensions (e.g. Cumulative Energy Demand CED [28] or Material Intensity Per Service Unit MIPS [29]). However, care needs to be taken when using screening indicators. It must be double-checked that all environmental aspects of the product life cycle that are relevant for the goal of the study are considered. This is typically not the case if a single screening indicator like CED or MIPS is used only.

Simplified LCAs can be conducted quicker but the usefulness of the results is limited by the simplifications. They can be used to identify hot spots and help setting system boundaries.

Insignificant factors or parts of the life cycle might be excluded based on this screening. Straightforward examples include the deletion of a life cycle part that is conserved between compared products. Other identified simplifications include focusing the LCA study on a subset of input or output data in Life Cycle Inventory and/or a limited number of impact categories that are of significance and importance for the goal of the study. A typical example is Carbon Footprint studies that only look at climate change aspects. However, these studies risk ignoring other environmental challenges potentially relevant for the considered products.

2.2.2 *Inventory*

2.2.2.1 Gathering Data

This phase establishes an inventory of all elementary flows of the system defined in the goal and scope definition normalized per functional unit. According to ISO 14044:2006 "elementary flows" are all materials, substances, energies, spaces etc.

that are directly taken from the environment or released directly back into the environment, so basically all input and output flows into and from the system to the environment. They represent the final Life Cycle Inventory result that will be translated in the Life Cycle Impact Assessment (LCIA) later on. In addition, ISO 14044:2006 covers intermediate flows that are used to connect processes within the system boundaries i.e. stay in the Technosphere. The elementary and intermediate flows are obtained for each unit process and then normalized to the unit process' output. The total amounts of the processes and their flows are quantified for the reference flow, which is often based on a linear, constant relationship.

LCAs are very data intensive, so data availability is the key challenge. It is necessary to gather all relevant inputs to and outputs from the activity. Relevant inputs and outputs are related to the function or product generated by all processes within the system boundaries. Data availability might be enhanced by existing databases. These often only have secondary data from industry averages for a certain or all regions and therefore do not contain primary data of the specific suppliers or end-of-life operators. Specific manufacturing data can be derived from environmental management systems, energy bills, and other financial inventories. However, the completeness of this data varies between different input flows that are usually only recorded for financial reasons (so not if free of charge) and output flows (that are often not covered especially if released free-of-charge and not regulated). Theoretical models and stoichiometric equations can support data gathering but either contribute to worst case assumptions, as some emissions are covered in emission control. Contrastingly, these models can also lead to best case assumptions as the efficiency of chemical reactions is usually not ideal. Thermodynamic calculations can support data gathering but only provide the minimum energy demand. It is clear that real data differs from modeled data. Data consistency is often an issue, as some data points are measured while others are only estimated, modeled/calculated. Measured data may also vary depending on measurement accuracy and representativeness of the measurement period (just individual samples versus constant measurements over multiple seasons).

Elementary flows occur between the system and environment, intermediate flows stay within the system. Data inconsistency occur between measured, calculated, specific, and average data.

Processes can also exhibit non-linear or dynamic behavior. This can be described by distribution factors. Distribution factors define how the process' input flows are distributed over the different output flows of a process. This means that the distribution factors define the share of the pre-products and raw materials in the targeted product and determine how much is lost as emissions (solid, liquid, airborne). These distribution factors can be constant or dynamic. They may be dynamic due to quality deterioration of input flows with an increasing recycling rate (Fig. 2.3) or due to higher energy consumption if the output of the process needs to be increased [30].

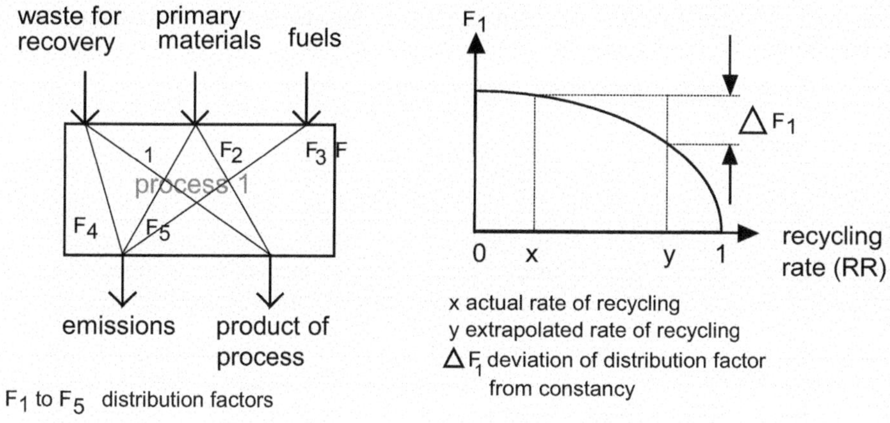

Fig. 2.3 Distribution factors between input and output process flows (left) and example for dependency on recycling rate [31]

For processes with non-constant distribution factors, the elementary flows are modeled depending on the influencing factors, leading to dynamic inventories.

> Non-linear distribution factors between process input and output flows are especially relevant in consequential LCAs where the non-linearity is caused by the investigated system change.

In addition, there are processes with co-products, processes that produce sequentially different products (e.g. in automotive assembly lines) or overhead input/output flows used for various processes that are only somewhat relevant for the functional unit. This leads to allocation questions when compiling the data for the LCA. For example, the amount of the yearly energy consumption of a production hall for lighting and heating must be distributed between processes within the system boundaries and other—unrelated—processes.

2.2.2.2 End-of-Life and Co-product Allocation Options

Different options for allocating the benefits and burdens of the end-of-life phase exist in terms of generation of secondary materials and energy. The same logic applies to processes with multiple co-products whose use extends beyond the assessed system. There is no general correct or incorrect approach for allocation. However, it is important to reflect on the chosen method when interpreting the results because results are impacted by the chosen allocation methodology. End-of-life allocation is a good example for the final numbers of an LCA—for example the difference between compared options or scenarios—being an artifact of the chosen methodology and assumptions.

Cut-Off Approaches

The cut-off approach is the simplest approach. It assumes that either (1) the elementary flows for the generation of secondary materials (E_{Sin}) used in the product are zero, (2) the elementary flows for the collection and treatment of waste streams (E_{Sout}) are zero or (3) both (1) and (2) are appropriate to a certain, varying degree:

- The first approach incentivizes the use of recycled content and allocates no environmental burdens or benefits to the production of secondary materials (E_{recycled}). The environmental impacts for the share of recycled material (R_1) are not considered while the share of virgin, primary materials ($1 - R_1$) with its up-stream environmental burdens or benefits is fully account for (E_V).

  However, this approach can lead to a situation in which—in a scrap constraint market like steel where all of the scrap is already being used—the wrong incentive is provided because increasing the recycled content simply takes away the opportunity for other producers to keep their recycled content at the previous level. Hence, there is no incremental environmental advantage from an overall perspective. Such an approach could even lead to results that disincentivize real transformations like the switch from conventional to green steel.
- The second approach incentivizes products with a high degree of recycling capability. This leads to design for recycling or design for disassembly approaches. A moderate approach to incentivize such design approaches is to not allocate any environmental burdens of waste collection and handling to the originating product system.

  An even higher incentive is provided if accounting for the avoided burden (or recycling credit). Here, the environmental impacts or elementary flows for the production (cradle-to-gate) of the substituted virgin material or product (E_{V2}) are subtracted from the environmental burdens of the waste generating system. This approach can also be used when comparing different end-of-life options.

  However, such an approach makes sense only where there is a high demand for the product of recycling and the product design actually supports a sustainable recycling approach. This can for example be reflected in the assumed average or measured recycled content factor for a subsequent system (R_2). However, the environmental burdens or benefits cannot be omitted for the portion that is unlikely to be used for the subsequent product system and needs to be disposed of (E_D).

> Cutting-off secondary materials' generation rewards recycled content but creates issues in constrained secondary markets; cutting-off end-of-life operation helps Design for Recycling.

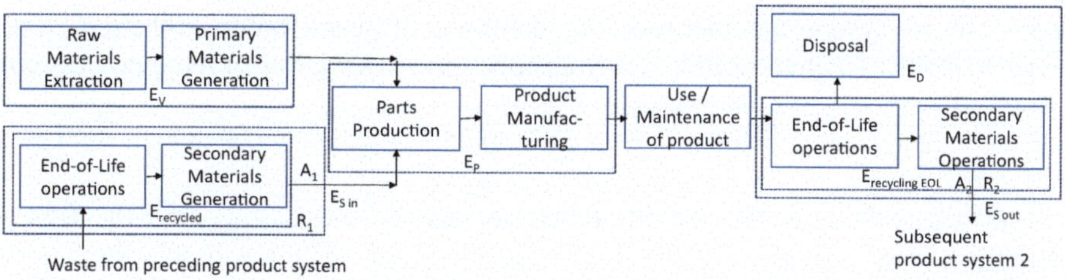

Fig. 2.4 Product system with in- and outgoing secondary material generation. ©*Wulf-Peter Schmidt 2024. All Rights Reserved*

- The third approach includes factors that determine how much of the environmental burdens or benefits are allocated to the assessed system. This allocation factor (A) can be 50% to ensure that receiving and giving systems are assigned the same share of burdens or benefits. Alternatively, the factor can reflect economic or quality aspects. Because quality is typically not one-dimensional, the choice of a quality allocation factor requires subjective weighting approaches. Similarly, economic factors are not straight-forward and may require subjective choices related to time and geography. Such choices need to be justified in line with the goal and scope definition.

Figure 2.4 shows an example for allocation approaches linked to material recycling. Analogous approaches are possible for reuse, remanufacturing, energy recovery, and other end-of-life options as well as processes with co-products.

The cut-off approaches impact on the system boundaries and may require an adjustment in the iterative approach of establishing an LCA. The cut-off approach is a simple way to separate different systems that are interlinked in a circular economy. The international system for Environmental Product Declarations (EPD)—according to ISO 14025—recommends a cut-off approach that defines the boundaries between the life cycles that are linked via recycling, reuse, or recovery products as the point where the material/part/ energy has its lowest market value. This is usually before the waste material/product/energy flow is collected for recycling/reuse/recovery. The recycling/reuse/recovery processes should be included in the EPD of the product only if the recycled material/reused product/recovered energy is used [32]. With this cut-off approach the environmental burdens or benefits of a product life cycle (E) are calculated using Eq. 2.1 where E_D are the environmental burdens or benefits of disposal:

$$E_P = (1 - R_1) \times E_V + R_1 \times E_{\text{recycled}} + (1 - R_2) \times E_D \tag{2.1}$$

The Eq. 2.1 basically means that the environmental burdens or benefits of a product P is the sum of the environmental burdens/benefits of the virgin E_V and recycled material E_{recycled} plus those for the end-of-life operation E_D. Each of these three factors is weighted by the share of virgin or actually recycled material. For example, Eq. 2.2 reflects the situation of a plastic bottle with 25% recycled content and a deposit return system that increases the likelihood of plastic bottles to be reused or recycled to 90% (including all losses from collection to recycling or reuse processes).

$$E_{\text{Plastic bottle}} = 0.75 \times E_V \times E_V + 0.25 \times E_{\text{recycled}} + 0.1 \times E_D \qquad (2.2)$$

The factors E_D as well as the difference between E_V and E_{recycled} drive the degree of the incentive provided for the use of recycled content in the base system 1 or the generation of secondary material for system 2. If the environmental burdens of disposal are low (e.g. for inert material) or if there are disposal benefits (e.g. in some cases of controlled handling of organic materials generating compost or recovered gasses/energy), there is a low incentive for improving design for recycling. The avoided burden approach that leads to a system expansion (Sect. 2.2.2.2.3) provides a more holistic perspective.

Circular Footprint Formula Approach

The Circular Footprint Formula (CFF) proposal of the EU Joint Research Center (JRC) [33] is part of a broader Product Environmental Footprint (PEF) methodology of the EU Commission that aspires to create a common European environmental assessment methodology for products and services based on LCA [34]. CFF distributes the environmental benefits and burdens among the supplier and user of recycled materials. It is calculated by adding up the environmental burdens of material production (cradle-to-gate, E_P), the environmental burdens or benefits related to the input of secondary material ($E_{S\,\text{in}}$), environmental burdens or benefits related to the output of secondary materials ($E_{S\,\text{out}}$) as well as the environmental burdens and benefits related to energy recovery (E_{ER}) and the environmental burdens of disposal (E_D).

$$\text{CFF} = E_P + E_{S\ \text{in}} + E_{S\ \text{out}} + E_{ER} + E_D \qquad (2.3)$$

$$E_P = (1 - R_1) \times E_V + R_1 \times E_{\text{recycled}} \qquad (2.4)$$

with R_1 = recycled content, E = Environmental burden or benefit, V = virgin material.

This is very similar to Eq. 2.1, however, CFF adds the factors ($E_{S\,\text{in}}$), ($E_{S\,\text{out}}$) and (E_{ER}) to introduce additional allocation factors specifically for secondary materials entering or leaving the studied system as well as to introduce the option energy recovery.

$$E_{S\ \text{in}} = -(1 - A) \times R_1 \times (E_{\text{recycled}} - E_V \times \frac{Q_{S\ \text{in}}}{Q_P}) \qquad (2.5)$$

A = allocation factor of burdens or benefits between supplier and user of recycled content (factor between 0 and 1).

$Q_{S\ in}/Q_P$ = Quality ratio of ingoing secondary material to primary material.

Please note that the term "$-(1-A) \times R_1 \times (E_{recycled})$" in Eq. 2.5 is only introduced to avoid the environmental impact of the generation of recycled content being accounted for twice. This term is subtracted again from the second half of Eq. 2.4. This is demonstrated in Eq. 2.6

$$-(1-A) \times R_1 \times (E_{recycled}) + R_1 \times E_{recycled} = A \times R_1 \times E_{recycled} \tag{2.6}$$

Applying Eq. 2.5 to the above example of 25% recycled content in the plastic bottle there might be a 20% lower quality of the recycled compared to the virgin material. This could be a lower impact strength, differences in color, etc. The allocation factor A could be set to 0 to nullify any environmental burdens related to generating the recycled plastic or to 1 to fully account for these environmental impacts. In Eq. 2.7 a 50% allocation is chosen what basically allocates only half of the environmental burdens of generating the recycled plastic to the studied system and leaving the other half outside the system boundaries (theoretically burdening another system that generates the original waste leading to recycled plastic).

$$E_{S\ in} = -0.5 \times 0.25 \times (E_{recycled} - E_V \times 0.2) = 0.025\,E_V - 0.125\,E_{recycled} \tag{2.7}$$

This example shows that the factor ($E_{S\,in}$) penalizes secondary materials of lower quality than primary, virgin materials by adding the extra-burden in this case 2.5% of the environmental impacts of the generation of virgin materials. The allocation factor can nullify ($A = 1$) or maximize ($A = 0$) this penalty.

The factor ($E_{S\,out}$) covers the secondary material generated out of the studied system's waste going into the next system:

$$E_{S\ out} = (1-A) \times R_2 \times \left(E_{recycling\ EOL} - E_{V2} \times \frac{Q_P - Q_{S\ out}}{Q_P}\right) \tag{2.8}$$

R_2 = the proportion of material in the product that will be recycled and/or reused in a subsequent system measured as the output of the recycling/reuse facility (i.e. considering losses in collection and recycling/reuse).

V_2 = assumed substituted virgin material in the subsequent system.

$Q_P - Q_{S\ out}/Q_P$ = Quality ratio of outgoing secondary material to primary material of the studied base system. Note: For explanatory reasons this factor is displayed differently than in the original proposal [33].

In the above example of the plastic bottle 90% of the used plastic bottles will be used in a subsequent system. This might happen by collection of the used bottles, grinding, and mechanical recycling to secondary material used in an automotive plastic part like an engine shield i.e. an unrelated, subsequent life cycle system. Assuming the quality of the recycled, secondary material in the engine shield might be the same as the primary, virgin material in the engine shield but 25% worse than in the plastic bottle (studied base system) the Eq. 2.8 leads to the following calculation based on the same allocation factor like used before:

$$\begin{aligned} E_{S\,\text{out}} &= 0.5 \times 0.9 \times \left(E_{\text{recycling EOL}} - E_{V2} \times 0.75\right) \\ &= 0.45 \times (E_{\text{recycling EOL}} - 0.675 \times E_{V2}) \end{aligned} \tag{2.9}$$

This example shows that the factor ($E_{S\,\text{out}}$) incentivizes a circular economy model that allows for a high actual reuse/recycling in a subsequent system at a high-quality level by allocating a recycling credit via the subtracted environmental burdens of the substituted virgin plastics. A so-called "down-cycling" to lower quality usages in subsequent systems is less incentivized by referring the quality factor to the quality requirements of the base system. Again, the allocation factor can nullify ($A = 1$) or maximize ($A = 0$) this incentive.

$$\begin{aligned} E_{\text{ER}} = (1 - B) \times R_3 \times \big(&E_{\text{ER}} - \text{LHV} \times X_{\text{ER, heat}} \times X_{\text{ER,heat}} \\ &\times E_{\text{SE,heat}} - \text{LHV} \times X_{\text{ER, elec}} \times E_{\text{SE, elec}}\big) \end{aligned} \tag{2.10}$$

with R_3 = energy recovery rate, ER = energy recovery, LHV = lower heating value of the material in the product used for energy recovery, SE = substituted energy, LVH = lower heating value of the material in the product used for energy recovery. B is the allocation factor of energy recovery processes (both for burdens and credits of energy recovery).

If in the example of the plastic bottle half of the not returned used plastics are going to an energy recovery (i.e. energy recovery rate of 5% ($R_3 = 0.05$) having in mind the 90% return rate) the environmental burdens of the energy recovery emissions (E_{ER}) are added. However, there is an energy recovery credit by subtracting the substituted emissions/impacts of generating heat and electricity. This works for plastic that has a positive (lower) heating value.

Finally, the disposal of the remaining volumes (in the above example 5% of the material) is accounted for.

$$E_D = (1 - R_2 - R_3) \times E_{\text{End-of-life Disposal}} \tag{2.11}$$

The Circular Footprint Formula (CFF) approach is affected by several factors or ratios that are (partially) subjective. It requires a transparent, well-justified, and fair method of quantifying these numbers. The allocation factor A should reflect market demand. However, market demand will evolve over time and varies between regions. The allocation factor B seems to be not really needed having in mind that the energy recovery rate should already reflect what portion of the material is going to an energy recovery. The quality factors are difficult to merge into single-factor quantitative indicators. The CFF may also require data for processes that are not used in real life, for example, the disposal might not exist [35].

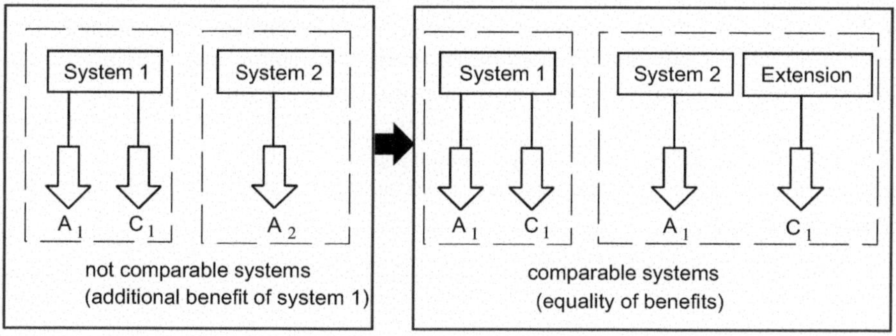

Fig. 2.5 Transforming non-comparable systems to comparable ones by enlarging the systems (compare [36]). ©*Wulf-Peter Schmidt 2024. All Rights Reserved*

The Circular Footprint Formula is based on multiple subjective factors that decide which end-of-life option is incentivized. Factors require justification and a sensitivity analysis.

System Expansion

The system extension approach avoids allocation factors altogether and therefore has a reduced subjectivity. Avoiding allocation factors is recommended in ISO 14044 by "expanding the product system to include the additional functions related to the co-products" of the compared system.

The underlying idea is that the fundamental assumption of comparative LCAs is the equality of all functions or benefits (services) of the compared systems, i.e. a comparison of environmental impacts is only acceptable and fair if the compared systems offer the same benefits (B) in terms of products or services [36]. If system 1 offers more benefits (products A and C) but is responsible for more environmental burdens while system 2 causes less benefits (only product A) and less environmental burdens a direct, unadjusted environmental comparison is impossible. For an environmental assessment, it must be determined whether the higher environmental burden of system 1 is mitigated by the additional benefit of the product C. Because of the subjectivity of benefits/values of products such a decision is questionable. For example, systems with different recycling/disposal options produce different benefits. To achieve comparability despite different benefits of the compared life cycles, the systems have to be extended to include the production of the respective lacking product (see Fig. 2.5).

In this example, system 1 generates product C_1 in addition to the main product A_1. However, the compared system 2 only generates product A_2. Therefore, these two systems are not directly comparable (Eq. 2.12), because only the quality and quantity of the products A_1 and A_2—and the benefits of these products—are assumed to be equal (Eq. 2.13). Comparability can be ensured by extending the system 2 by the relevant parts of the life cycle generating the complementary good (C_C). If the quality and quantity of the products C_1 and C_c are equal (Eq. 2.14), both systems generate equal benefits (Eq. 2.15). These complementary goods must be produced from virgin, primary raw materials, because there are no secondary materials available within the system boundaries.

$$A_1 \cup C_1 \neq A_2 \tag{2.12}$$

$$A_1 = A_2 \tag{2.13}$$

$$C_1 = C_C \tag{2.14}$$

$$A_1 \cup C_1 = A_2 \cup C_C \tag{2.15}$$

One difficulty of this approach is identifying all benefits and determining the method of generating representative, complementary products. For example, the choice of renewable versus coal-based electricity for the complementary product's electricity needs matters significantly. It is crucial to include these subjective choices in the conclusion of an LCA because they have the potential to fundamentally change the determination of the superior system from an environmental perspective.

System extension ensures system comparability by adding processes to generate lacking, complementary products. The choice of complementary processes requires justification.

The importance of system extension can be shown by comparing virgin and recycled paper. Virgin paper has lower environmental burdens than recycled paper if there is no system expansion and the comparison is based on different benefits for the compared systems [37].

In Fig. 2.6, all systems generate the same quantity (and quality) of paper (P) if the quality of recycled paper is assumed to be equivalent to virgin paper within the definition of the functional unit. However, the energy produced by waste paper energy recovery (E) and the forest products N (biodiversity, beauty, humus, carbon removal) are different.

Because energy is generated out of non-recycled waste paper, the quantity of energy (E) decreases with an increasing recycling rate, i.e. the energy product is different between the three systems in Fig. 2.6 (Eq. 2.16). This simplification only considers the energy generation and not the energy consumption of the system. In

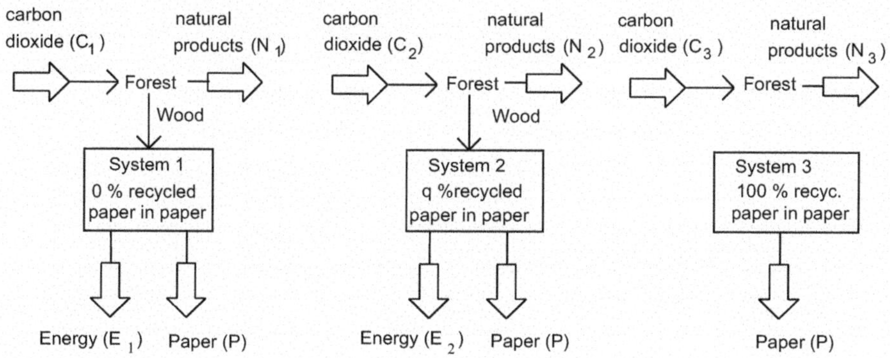

Fig. 2.6 Comparison of virgin and recycled paper without system expansion leading to different system benefits [31]

these simplified recycling scenarios, only the paper incineration—and not landfill—is considered as end-of-life option [36].

Looking at the other products of the systems, the products of the forest must be examined. Forest products are generated without human influence and offer an immediate benefit that differs from abiotic resources in the earth's crust that do not provide an immediate function or benefit nor generate products without human interference. This difference is important because a reduction of consumption of abiotic resources has no immediate product/service benefit as a consequence. The preservation of abiotic resources for the future is no benefit in itself, if benefit is defined by goods or products. It only improves opportunity costs, i.e. keeps options for future generations which is beyond the scope of the LCA. The amount of forest products increases with the increasing recycling rate (Eq. 2.17).

The absorption and storing (in humus, undergrowth, or paper) of carbon dioxide by forests varies between all systems and is usually seen more as an environmental advantage rather than a product/good-based benefit and could be alternatively covered in Life Cycle Impact Assessment (LCIA) without the need of system extension.

Comparisons between systems that consume versus preserve natural resources need to reflect on the co-benefits of preserving nature by system extension or in LCIA.

In this comparison, the three systems ($x = 1, 2, 3$) produce the same amount (and quality) of paper but require different amounts of energy and natural products (Eq. 2.18). This leads to different product/service-based benefits of the three systems (Eq. 2.19). It cannot be decided whether different environmental performances are compensated by differences in benefits.

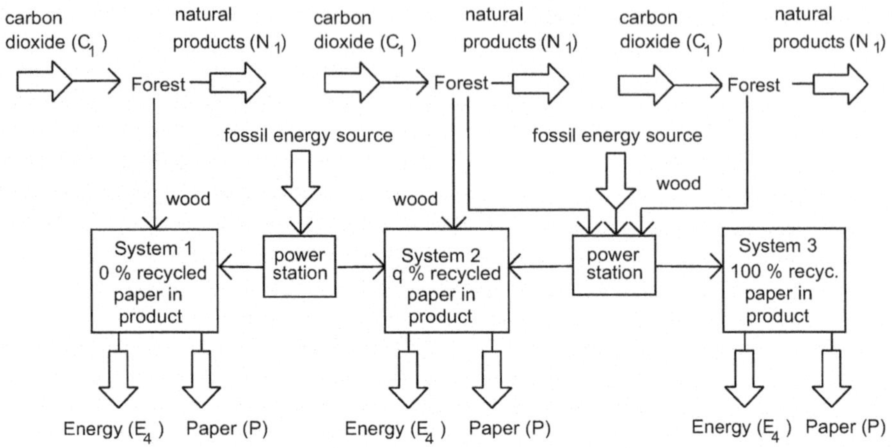

Fig. 2.7 Comparison of virgin and recycled paper with system expansion leading to comparable system benefits [31]

$$E_1 > E_2 > E_3 = 0 \tag{2.16}$$

$$N_1 < N_2 < N_3 \tag{2.17}$$

$$B_x = P \cup E_x \cup N_x \tag{2.18}$$

$$B_1 \neq B_2 \neq B_3 \neq B_1 \tag{2.19}$$

The difficulty of system expansion lies in the justified choice of processes generating the complementary products. Figure 2.7 shows a system expansion leading to comparable system benefits (Eq. 2.20). This is done by reducing the natural products to the amount in the system without any recycled paper because there is no complementary process to generate the natural products at the same quality. Energy, on the other hand, can be produced for example by adding power stations. This suggests that a fair comparison is now guaranteed.

$$B_1 = B_2 = B_3 = P \cup E_4 \cup N_1 \tag{2.20}$$

However, there are new (subjective) choices that can influence the final result. For example, one must determine the source of the complementary energy (whether fossil or renewable, what efficiency of energy production).

The final environmental impact and the conclusion of whether virgin or recycled paper is environmentally preferable depends on the methodological approach chosen. A narrow system boundary and definition of functional unit that does not consider the benefits of forest products leads to the conclusion that virgin paper is preferable

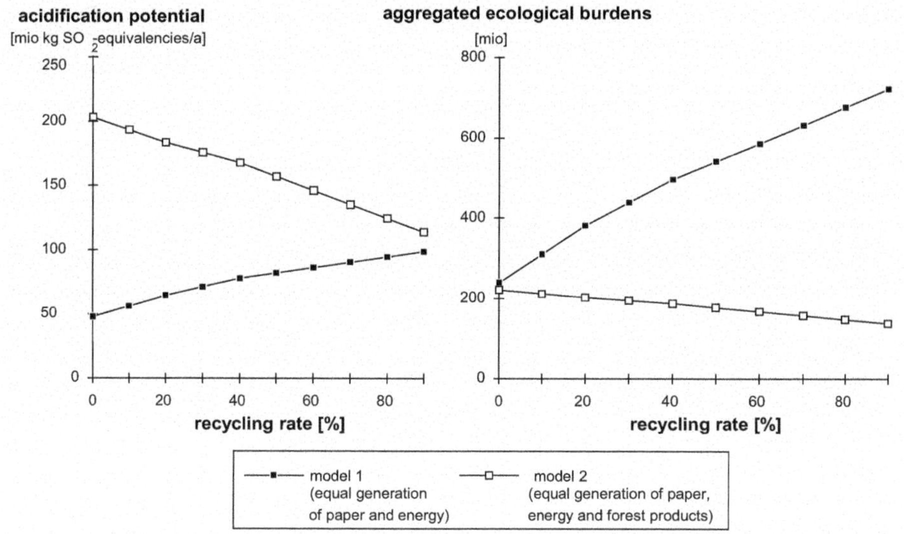

Fig. 2.8 Acidification potential and aggregated ecological burdens of paper consumption in 8 European countries using an LCA without system expansion (model 1; see Fig. 2.6) and with system expansion (model 2, see Fig. 2.7) [31]. ©*Wulf-Peter Schmidt 2024. All Rights Reserved*

environmentally. An expanded system boundary and a definition of a functional unit that ensures the same benefits including forest products leads to the conclusion that recycled paper is environmentally preferable, so higher the recycling rate the better (at least until the increased recycling rate does not lead to additional issues as laid out Sect. 8.5 and Fig. 8.5) (Fig. 2.8).

Overview of Other Methods

The choice of an allocation method for co-products as well as end-of-life processes is a contested aspect and different assessments of these methods have been offered. Various sub-methods and further methods exist. Table 2.1 provides an overview of various allocation methods and an assessment of these methods.

2.2.2.3 Data Quality

Data uncertainty covers the inaccuracy and lack of representativeness of data. An appropriate quality of data is required for LCAs. Data quality in LCAs is defined according to ISO14040:2006 as "characteristics of data that relate to their ability to satisfy stated requirements". These characteristics include validity, relevance of data for study (fit of geography, time, technology, etc.; specific or average data), appropriate measurement methods, completeness satisfactory for the needs of the study,

Table 2.1 Assessment of allocation methods by [32] with additions

	Easy to use	Readily available data	Generalizable results	Reflects decisive characteristics	Life cycle scope	Explicit, justified, and evaluated	Comprehendible	Relevant to decision makers	Legitimate	Reproduceable
Simple cut-off	+	+	+	–	+	+	+	+	+/–	+
Cut-off with economic allocation	+	+	+	–	+	+	+	+	+/–	+/–
Cut-off plus credits	+/–	+/–	–	+/–	+/–	+/–	+	+	+	–
Allocation to material losses	+	+	+	+/–	+	+	+	+/–	+/–	+/–
Allocation to virgin material use	+	+/–	+	+/–	+	–	+	+/–	–	+/–
50/50 methods	+	+/–	+	+/–	+	+	+	+	+	+/–
Quality adjusted 50/50 methods	–	–	+	+	+/–	+/–	–	+	+	+/–
Circular Footprint Formula	–	+/–	+	+	+	+	–	+	+	+/–
Market price-based allocation	+/–	+/–	+	+	+	+	+	+	+/–	+/–

(continued)

Table 2.1 (continued)

	Easy to use	Readily available data	Generalizable results	Reflects decisive characteristics	Life cycle scope	Explicit, justified, and evaluated	Comprehendible	Relevant to decision makers	Legitimate	Reproduceable
Market price-based substitution	–	–	+	+	+	+/–	–	+	+	–
Price-elasticity approaches	–	–	+	+	+	–	–	+	+	–
Allocation at the point of substitution	–	–	+	+/–	+	+	–	+	+/–	–
System expansion	–	–	+	+	+	+	+	+	+	+

©*Wulf-Peter Schmidt 2024. All Rights Reserved*

documentation of data, uncertainty and precision, homogeneity, consistency and reproducibility. Quantitative and qualitative Data Quality Indicators (DQIs) describe these various aspects of data quality [38, 39].

However, data quality is not absolute and can depend on the way data is used/ applied in an LCA. The application-dependency of data quality was investigated recently in a United Nations GLAD (Global Life Cycle Access to Data) project. GLAD distinguishes between data quality at the point of creation and time of use of a dataset. In addition, data quality can be extended to include modeling details that usually differ between the various application contexts [40].

Different methods exist to assess whether life cycle data is appropriate for the intended application. These include the so-called "pedigree matrix approach" used by ecoinvent [41] or the data quality ranking system used by the International Reference Life Cycle Data System (ILCD) [42] and the qualitative pass/fail method used by the USDA LCA Commons, a collection of data repositories in the US [43]. The pedigree matrix provides indicator scores from 1 (best) to 5 (worst) for reliability, completeness, temporal correlation, geographical correlation, and further technological correlation. An uncertainty factor is calculated from this. The ILCD DQA method consists of six indicators, is applied to the process level, and also has scores ranging from 1 (best) to 5 (worst), depending on compliance with a set of pre-defined standards for each of the six indicators. The qualitative pass or fail check used in the USDA LCA Commons includes seven indicators to address nine out of ten DQCs (excluding consistency) as defined by ISO 14040:2006 [39].

> Data quality depends on factors inherent to the data itself & on the application/ use of data in LCAs. Data Quality Indicators, pedigree, rankings or pass/fail methods describe data quality.

Irrespective of the data quality assessment method used, it is important to work continuously on improving data quality of LCAs and to reflect on what conclusions can be drawn from the existing data quality and the final LCA results, i.e. the data uncertainty needs to be thoroughly reflected in the interpretation phase (see Sect. 2.2.4).

2.2.3 Impact Assessment

In Impact Assessments, the Life Cycle Inventory results of all elementary flows (resources, emissions, etc.) are assessed for their environmental impact. Elementary flows must be correlated to the relevant impact categories and suitable indicators. This is ideally done and justified in the goal and scope definition but might be refined in the iterative nature of LCAs. The inventory data is assigned to the corresponding impact categories (so-called classification) and the impact category indicator (CI) is calculated (so-called characterization) following the general formula below:

$$\mathrm{CI} = \sum_{s}\sum_{i} \text{Characterization Factor}(s,i,t) \times \text{Elementary Flow}(s,i) \qquad (2.21)$$

where s denotes substance, i is the region of an elementary flow, and t is the time over which the effect of elementary flows on impacts is considered. Most LCAs do not have inventory data available specific to the location in which the elementary flow occurs. This is a limitation of LCAs for all non-global impact categories.

The characterization factor converts the elementary flow to its equivalent impact based on its importance for the impact category. For example, the impact category climate change is measured by the indicator Global Warming Potential (GWP) in terms of CO_2-equivalency. The characterization factor converts emissions in terms of differences in the effectiveness of global warming: If 1 kg of methane has the same effectiveness or potential for global warming as 28 kg of CO_2, the characterization factor of methane is 28.

Local impact categories are not suitable for most LCAs because they require Life Cycle Inventory data specific to location, time, and even exposure of emission. The ideal formula to calculate such local impact category indicators is as follows:

$$\text{Characterization Factor}(s,i,t) = \sum_{j} \frac{\text{Effect}(s,j,t)}{\text{Elementary flow}(s,i)}$$
$$= \sum_{j} \left(\frac{\text{Fate}(s,j,t)}{\text{Elementary flow}(s,i)}\right) \cdot \left(\frac{\text{Exposure}(s,j,t)}{\text{Fate}(s,j,t)}\right) \cdot \left(\frac{\text{Effect}(s,j,t)}{\text{Exposure}(s,j,t)}\right) \qquad (2.22)$$

where j denotes the location of the receiver of emissions.

The impact of each elementary flow depends on local and time aspects for non-global categories that are often not covered in inventory. This limits the validity of these categories.

Some impact categories are considered in almost all LCAs (like Global Warming) while others are only chosen where seen as exceptionally important for the study (like land-use-related impact categories). The methodologies for modeling the chosen impact categories are typically either

- Endpoint related, i.e. following the cause-effect chain from elementary flows to the final damages caused to an "Area of Protection" such as human health, biodiversity, or natural resources. Due to the length of cause-effect chains, these methodologies are often complex and based on many assumptions, uncertainties, and value judgements. An example of an endpoint indicator linked to human health is the disability-adjusted life years (DALYs) that represent the years of life lost or impaired due to disease or injury.

- Midpoint related, i.e. following the cause-effect chain from elementary flows to changes in the natural environment without capturing the full consequences for the "Areas of Protection". An example of a midpoint indicator is the Global Warming Potential (GWP) expressed in CO_2-equivalent emissions. This indicator does not capture the implications of greenhouse gasses for human health and biodiversity but aggregates all greenhouse gasses into one number by taking into account their efficacy of global warming relative to CO_2 (or more precisely: it is their "time-integrated radiative forcing due to a pulse emission of a given component, relative to a pulse emission of an equal mass of CO_2" [44]).

Various impact categories and indicators are proposed [45–47], including those listed in Table 2.2; the sequence in the table indicates the level of maturity and acceptance in decreasing order.

In the next, optional step the impact category indicators can be normalized by dividing them by the total impact for each indicator in the reference region. This step puts the results of the study into perspective. This is done to make the quantitative values of all indicators more comparable. The normalized results can be used for the following, also optional steps of grouping and weighting.

Grouping is a qualitative, or semi-quantitative sorting and/or ranking of results across impact categories. Grouping may result in a ranking of impact categories with respect to their importance on a societal level. Such a ranking can then be combined with results from normalization to draw conclusions on the comparative importance of different impact categories in the product system under study [45].

Weighting quantitatively converts the different indicator results (or the results from normalization) with the help of (typically linear) weighting factors to a single number:

$$\mathrm{EI} = \sum_k w_k \times \mathrm{CI}_k \text{ or } \mathrm{EI} = \sum_k w_k \times \frac{\mathrm{CI}_k}{N_k} \tag{2.23}$$

With EI being the total, weighted potential environmental impact, w_k the weighting factor of the Category Indicator CI for the impact category k and N being the normalization factor.

Common examples for such methods include the distance-to-target approach. Distance-to-target (DtT) weighting is based on the distance of (desired) policy targets to current environmental situations, and recent methodological DtT developments are based on a weighting perspective for a single region or country, mainly considering environmental situations in consuming countries or regions [49]. The different types of weighting factors are very controversial.

In any case, weighting should not be used for comparative assertions disclosed to the public [50]. The reasons for this include the subjective, value-based nature of weighting, the sensitivity to anthropocentric, temporal, socio-economic, and spatial bias as well as the difference of such weightings between regions on a public but also

Table 2.2 Impact categories suggested for product/organization environmental footprints [47] and comments

Impact category	Impact category indicator (unit of measure)	Description	Comments
Climate change	Radiative forcing as global warming potential—GWP100 (kg CO_2 eq)	Increase in the average global temperature resulting from greenhouse gas emissions (GHG)	Most accepted; important to always use latest agreed upon factors of IPPC [48]. Also used in Carbon Footprint studies
Ozone depletion	Ozone Depletion Potential—ODP (kg CFC-11eq)	Depletion of the stratospheric ozone layer that protects from hazardous ultraviolet radiation	Well accepted; importance declining due to success of Montreal Protocol (see Sect. 1.1)
Acidification	Accumulated Exceedance—AE (mol H+ eq)	Acidification from air, water, and soil emissions (primarily sulfur compounds)	Common impact category in most LCAs; various other indicators exist including kg SO_2 eq.; regional importance
Photochemical ozone formation, human health	Tropospheric ozone concentration increase (kg NMVOC eq)	Potential of harmful tropospheric ozone formation ("summer smog") from air emissions	Common impact category in most LCAs; regional and time bound importance
Eutrophication, terrestrial	Accumulated Exceedance—AE (mol N eq)	Eutrophication and potential impact on ecosystems caused by nitrogen and phosphorous emissions	Not all LCA studies distinguish between the different types (also because inventory may not distinguish what type of water is polluted); partially different indicators used
Eutrophication, systems freshwater	Fraction of nutrients reaching freshwater end compartment (kg P eq)		
Eutrophication, marine	Fraction of nutrients reaching marine end compartment (kg N eq)		

(continued)

Table 2.2 (continued)

Impact category	Impact category indicator (unit of measure)	Description	Comments
Land use	Soil quality index, representing the aggregated impact of land use on biotic production, erosion resistance, mechanical filtration, ground-water replenishment (Dimensionless—pt)	Transformation and use of land for agriculture, roads, housing, mining etc. The impact can include loss of species, organic matter, soil, filtration capacity, permeability	A multitude of methods exist, evolving methodologies, limited data availability, and data quality for land use thus additional caution required in interpretation
Water use	Weighted user deprivation potential (m^3 world eq)	Depletion of available water depending on local water scarcity and water needs for human activities and ecosystem integrity	Several methods exist—including distinguishing between types of water; limited data availability and quality thus additional caution required during interpretation
Resource use, minerals, and metals	Abiotic resource depletion—ADP ultimate reserves (kg Sb eq)	Depletion of non-renewable resources and deprivation for future generations	Various methods exist; doubts whether this is an environmental issue at all or more of an economic one; back-stop alternatives ignored; risk of double-counting (GWP-ADP fossil)
Resource use, fossils	Abiotic resource depletion, fossil fuels—ADP-fossil (MJ)		

(continued)

Table 2.2 (continued)

Impact category	Impact category indicator (unit of measure)	Description	Comments
Ionizing radiation, human health	Human exposure efficiency relative to U-235 (kBq U-235 eq)	Impact of exposure to ionizing radiations on human health	Various methods exist; doubts whether LCA is the right methodology to address toxic or radioactive impacts because of the aggregation of emissions over the life cycle while the real potential is much more linked to a dose-effect relationship than other impact categories (background emission levels, actual distance between emission and humans/animals, etc.); see also Eq. 2.22
Particulate matter	Impact on human health (disease incidence)	Impact of particulate matter emissions + its precursors (e.g. sulfur and nitrogen oxides) on human health	
Human toxicity, cancer	Comparative Toxic Unit for humans (CTUh)	Impact of absorbing substances into the body through air, water, and soil on human health; no direct effects of products on humans	
Human toxicity, non-cancer	Comparative Toxic Unit for humans (CTUh)		
Ecotoxicity, freshwater, etc.	Comparative Toxic Unit for ecosystems (CTUe)	Impact of toxic substances on fresh-water ecosystems	

©Wulf-Peter Schmidt 2024. All Rights Reserved

expert and scientist level due to different legal, exposure-related, cultural and socio-political backgrounds [51]. Similarly, weighting based on monetarization methodologies is linked to bias and subjectivity (see Table 2.7). Existing linear end-point weighting (and normalization) methodologies also have the tendency to cause a single impact category dominating result [52]. The resulting single-score or end-point score falsely implies a clarity of results by hiding trade-offs. This approach is risky not only for external but also for internal use of LCAs. In fact, such an approach may transfer a decision on the preferred option from the decision maker to the LCA expert. This is inappropriate, as LCAs are a decision support tool and not a decision machine. To support good decisions, it is important to show trade-offs also from an environmental or sustainability perspective transparently.

Normalization, group, ranking and weighting of life cycle impact indicator results are optional steps. ISO does not support weighting for comparative assertions disclosed to the public.

The opposing argument is that some weighting methods may actually help to make decisions more straight forward if they derive weighting factors from democratically legitimized policy targets of the country to be considered (e.g. Distance-to-target methods). This may strengthen the acceptance of the method among different stakeholders [49].

2.2.4 *Interpretation*

The interpretation of the results of inventory (LCI) and impact assessments (LCIA) is the last phase in the iterative LCA methodology. According to ISO 14044:2006, results should be interpreted by the following steps (see Fig. 2.9):

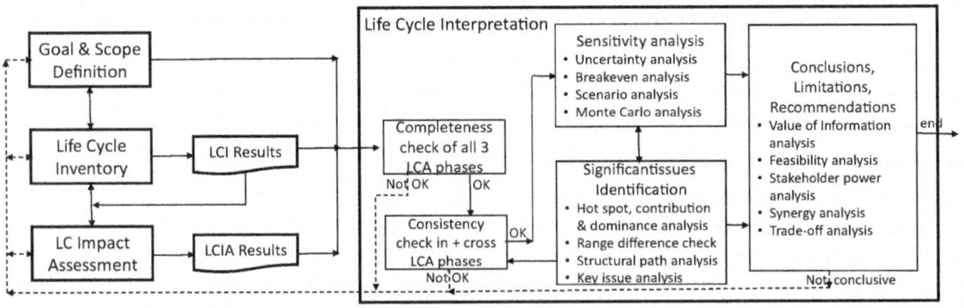

Fig. 2.9 Five recommended steps for Life Cycle Interpretation (based on [53]). ©*Wulf-Peter Schmidt 2024. All Rights Reserved*

- Completeness checks—the completeness of data, system boundaries, LCIA indicators etc. are checked against the defined cut-off limits (e.g. coverage of 95%).
- Consistency checks—consistency needs to be ensured across all phases of an LCA, including alignment of LCI, LCIA, and interpretation with goal & scope as well as between each other in terms of system boundaries, methods, data quality, etc. Identified shortfalls should ideally be corrected before proceeding.
- Analysis of sensitivities—given the inherent uncertainties and/or biases in models, assumption, and data, the sensitivity of the LCI and LCIA results to changes must be investigated. The impact of changing assumptions, simplifying data, and/or the totality of all uncertainties is calculated or a Monte Carlos analysis is performed. Break-even points are identified, e.g. based on what assumption change flips the preferred system from A to B. All this aids interpretation by identifying how robust results are.
- Identification of the significant issues extrapolated from the results of the LCI and LCIA phases—based on hot spot, contribution, dominance, and other analysis to ensure the final conclusions cover the important points.
- Provision of conclusions, limitations, and recommendations to the decision-makers and stakeholders after all checks and analysis suggest the LCA study is conclusive with respect to the goal definition. The recommendations should be made—following a feasibility check—to those stakeholders that have been identified in the stakeholder power analysis as being in a position to take the recommendations on board, bearing in mind which stakeholders may benefit from the recommendations. Trade-offs and synergies should be transparently communicated. If the uncertainties require additional investigation, a value of information analysis may help to focus such an investigation onto a few key data points that impact the uncertainty of the decision the most.

Crucially, the impact of the chosen goal and scope definition as well as final numbers of the Life Cycle Inventory and Life Cycle (Impact) Assessment must be critically reflected upon, taking into consideration for example:

- What the impact of the different assumptions, simplifications, and methodological choices (like system boundaries, functional unit, allocation rules, chosen LCIA methods) is. The impact of data uncertainties or gaps on the final LCI and LCIA results should also be considered, i.e. are these results just an artifact of these choices or study limitations rather than reflecting real-life?
- What difference between the assessed options is significant enough to conclude that one option is superior or inferior to other(s)?
- How to deal with trade-offs?

The effects of assumptions, simplifications, and methodological choices are best reflected in a scenario assessment where their impact can be demonstrated—for example by showing how key results of Life Cycle Impact Indicators change when best/worst case assumptions are included. It is important to derive a realistic range of results. In fact, given the inherent variation of complex life cycles, one final number as the result of a LCA indicator is unrealistic. All data has ranges because it varies within industry (secondary data) and/or over time (e.g. primary data). The life cycle system boundaries always represent simplifications due to cut-off criteria. In addition, various assumptions and data uncertainties exist. Therefore, it is appropriate to reflect the range including all uncertainties and scenarios. It is even more essential to reflect on the range and not on the average result especially for comparative LCAs because the definitive conclusion that A is better than B is only justified if there is a sufficiently significant difference between the ranges.

> A thorough LCA interpretation includes various critical checks & analyses that investigate the full ranges of results rather than jumping to conclusions on simplified one-dimensional results.

A third-party expert (panel) review is usually recommended for published LCAs.

2.3 Life Cycle Costing (LCC)

2.3.1 Define a Goal, Scope, and Functional Unit

Like LCAs, the goal, scope, and functional unit of an LCC study must be defined as a first step. Different to LCAs also the perspective of the LCC should be defined i.e. whether costs are captured from consumer, producer, or societal/governmental perspective [54]. It is helpful to define the types and categories of costs covered as well as mention whether a "discounting" (i.e. lower weighting) of future costs is part of the study.

In this step, the type of LCC must be defined because this influences several aspects of an LCC. There are fundamental differences between conventional, societal, and environmental LCC studies in terms of system boundaries, intention/perspective, and cost categories covered. External costs might be included, only partially included or not included at all (see Table 2.3).

A "conventional LCC" usually considers the perspective of the producer or customer alone, only covering costs that are immediately tangible and result from one of these life cycle actors. Future costs are typically discounted based on expectations of market interest rates and inflation. The system boundaries of conventional LCC studies are often narrower than for LCAs. For example, in conventional LCAs the market price of a product represents all up-stream costs and/or the end-of-life phase is neglected if there is no direct cost to the producer or customer. Such studies may confine the LCC to a Total Cost of Ownership (TCO) methodology that is not in line with the LCC ambition in term of scope.

Table 2.3 Differences between conventional, environmental, and societal LCC [56]

Aspect	Conventional LCC	Societal LCC	Environmental LCC
System boundaries	No EoL phase if not borne by the main actor, upstream life cycle simplified (=price)	Whole life cycle	Whole life cycle
Perspectives: Actors	Mainly manufacturer, user/consumer	Mainly government/ society	Mainly manufacturer, user/consumer
Reference unit	System, item, product	System	Functional Unit like in LCAs
Cost categories	Mainly acquisition costs (R&D costs, investment costs) and ownership costs (operating costs, maybe disposal costs)	Mainly costs of construction, maintenance, and environmental damages; taxes/ subsidies have no net cost effect	Mainly costs of development, materials, energy, machines, labor, waste management, emission controls, transports, maintenance/repair, liability, taxes, subsidies
Externalities	Internal costs	Internal plus all externalities	Internal plus external costs expected to be internalized if not covered in LCA
Discounting of LCC result	Recommended	Recommended	Inconsistent and not recommended
Discounting of cash flows for calculation	Recommended	Recommended	Recommended
LCA according to ISO14040 series	No	Not recommended, with the risk of double-counting and inconsistencies	Yes
Standards	Various (ISO, IEC, SAE, AS/NZS, etc.)	Various, including United Nations and OECD	None (LCA: ISO 14040 series)
Application cases	Mostly internally within private organizations/supply chains	Mostly internally in a public organization	Mostly internal, but also for external communication

Total Cost of Ownership "is a form of cost analysis that aims to uncover all the likely costs that follow from owning certain kinds of assets or certain acquisitions during ownership life (and) attempts to reveal and summarize ownership costs that are obvious and known before purchase, but also the full range of so-called "hidden costs" which are easy to overlook" [55]. The ownership period can cover the service life (years the product/acquisition is in service), economic life (years the asset returns more value to owners than it costs to own, maintain, and operate), accounting life (years the asset/product impacts the organization's account) or the depreciable life (years over which owners charge depreciation expenses for an asset). In this sense, the ownership period starts when the product/acquisition begins to cause costs and ends when there are no ownership costs anymore. The TCO methodology is composed of six steps including the definition of life span and identification of cost categories (purchase price, maintenance, operating, financing costs etc.) [55].

Conventional LCCs reflect the cost for the producer or—as "cost of ownership"—consumer. Societal LCCs have a macro-economic perspective. Environmental LCCs complement LCAs.

"Societal LCCs" have extended system boundaries, broader perspectives and consider a larger set of cost categories. Societal LCCs go beyond the perspective of the producer and customer by taking a societal perspective (like in macro-economic cost-benefit analysis) and examine government and other public bodies as well. Therefore, transfer payments such as subsidies and taxes are excluded because they are internal to the system. Societal LCCs appraise external environmental and social costs and may also study opportunity costs. They can also be used as a stand-alone approach without environmental and social LCAs. The time period studied is broader than in conventional LCCs. Societal LCCs have a different time preference to conventional LCAs if discounting is applied (below 0.1% per year) [56].

"Environmental LCCs" have system boundaries and functional units equivalent to LCAs. Double-counting of externalities or environmental impacts in the LCC and the accompanying LCA should be avoided, i.e. externalities already fully covered in the accompanying LCA must not be covered in the environmental LCC. Like conventional LCCs, discounting only applies to the cash flow but not to other externalities covered [56].

2.3.2 *Inventory Costs*

All kinds of data—especially regarding externalities—may not be available and scenario development, forecasting, or other estimation methods may have to be employed. A general issue in data gathering arises during the allocation of costs when several products are produced simultaneously from a common input or process. For example, a refinery commonly produces about a hundred different products among which the costs of refinery heating must be distributed. In LCCs, the costs of personnel, capital, of acquired goods and services etc. must be allocated to the different products involved. Different cost allocation methods like the Relative Sales Value Method exist. In this approach, the total cost of producing multiple products is distributed between the individual products based on their relative sale values. At the point where a product leaves the production process as a finished product (the so-called "split-off point"), all costs upstream of this point must be distributed between the product and all other products still to be produced based on the share in proceeds. This adjusts the LCC data.

For internal costs, cost estimation or cut-off criteria can also be an option. Internal costs are those that a life cycle actor directly pays in monetary terms.

This is different to so-called external costs. The concept of external costs in LCCs goes back to at least 1918 when Pigou suggested a tax on environmental pollution to compensate for the difference between marginal private and marginal social interests in case of a negative externality [57, 58]. Negative externalities, or external costs, are costs that are not covered by the price of the product or economic activity in question. External costs can occur on a micro- or macro level and may also have indirect profit impacts (Table 2.4). An example is the negative impacts in terms of environmental or societal damages of environmental pollution of production that are not covered in the product price (see also Sect. 1.6) [59]. There are some regions that already internalize some of these environmental costs—in particular in the case of greenhouse gas emissions that are monetarized by an Emission Trading Scheme (ETS) or via a tax. Examples of taxes include vehicle tax systems depending on the CO_2 emissions of vehicles, varying from small taxes like in Germany with single-digit Euro per g CO_2/km to several ten thousand Euros for high CO_2 polluting vehicles in The Netherlands. However, this integration of "externalities" in areas without existing monetarization approaches is very difficult, biased, purely anthropocentric, and due to a macro-economic approach not in line with the usual micro-economic perspective of economic operators (customer, enterprises, business units) that focus on direct costs/real cash flow [58–60].

Internal costs are based on paid, estimated, and/or allocated current cost flows. External costs are not covered in the price of a product or activity.

Some external benefits do not benefit those who pay for the economic activity. Instead, they benefit third parties that do not compensate for these benefits—examples include various free-rider cases.

Different methodologies to gather external cost exist including:

- Willingness-to-pay based on public surveys (how much would you be ready to pay for a lower external impact?).

Table 2.4 Different costs categories covered; see also [15]

	Planet	People	Profit/prosperity
Externalities (costs, benefits)	Cost of biodiversity loss due to pollution etc.	E.g. Human well-being losses from pollution	E.g. reduced crop yield
Private costs or benefits	E.g. cost for pollution prevention	E.g. health & safety expenditures	E.g. raw material cost, interest rate

- Willingness-to-accept based on public surveys (how much money would you need to get at a minimum to accept the higher external impact?).
- Differences in market prices for goods that are affected by the external effect (e.g. real estate differences that can be attributed to a specific external impacts).
- Analyzing the costs of adaptation behaviors to escape the external effect (e.g. distance people are ready to travel to avoid external effect or cost they are paying for noise protecting windows).
- Analyzing actual repair/recovery costs or insurance costs.
- Analyzing actual costs that can be attributed to the external impact (e.g. lower harvest yield).
- Analyzing prevention costs (abatement costs to avoid the external impact in the first place).
- Market prices of e.g. emission certificates (e.g. EU Emission Trading Scheme/ ETS [61]).

Each of these methods will lead to different results (see also Table 2.7). Even within a methodology there is a range of potential monetized values for external effects. Each study is also only valid for a defined region, time, etc. For example, based on a meta-study of willingness-to-pay data for the value of human life, the OECD suggested a monetized value range of 1.5–4.5 million US$ with a base value of 3 m US$ for OECD countries. The monetized value range is between 1.8 and 5.4 m US$ for EU-27 with a base value of 3.6 m US$ in 2005 terms [62]. There are on-going attempts to develop a global impact measurement and valuation (IMV) standard for monetizing and disclosing positive and negative externalities [63].

The uncertainty in LCC data differs between internal (lower uncertainty) and external costs (higher uncertainty). Additionally, the uncertainty depends on when data is gathered and is higher if the study is based on forecasting (see Table 2.5).

LCC data uncertainty is driven by different methods for deriving costs (e.g. for externalities) and by the time and geographical distance between data gathering & the occurrence of costs.

2.3.3 Aggregate Cost by Cost Categories and Discounting

In the next step, all gathered costs must be aggregated. This can be done by dividing the costs into four different categories, namely economic costs, life cycle stages, activity types, and cost elements (see Table 2.6). Costs are best distributed into these categories sequentially when establishing an LCC analysis. The third and fourth levels are the most important. However, these are also the most difficult to systemically describe, as so many different facets must be considered. Budget cost and

Table 2.5 Uncertainties of different LCC elements along the life cycle for products with long development and use phases [58]

Cost Info (right) being in LC stage (down)	Product price	Taxation	Operation cost (energy)	Maintenance/ warranty costs	Insurance costs	Residual value	End-of-Life (EoL) cost/revenues	Uncertainty
Product Development Kick-off	Cost Target/ Marketing Price Assumption	Forecast	Engineering target and cost forecast	Target	Target (insurance classes)/ Insurance forecast	Target	Unknown/Target/ Derived from Predecessor	Highest
Product Development (design & verification)	Cost tracking (control model/ average)/ Marketing price assumption	Forecast	Cost tracking (control model/ average)/energy price assumption	Engineering actions	Engineering actions (insurance classes)/ Insurance forecast	Forecasting tools	Based on material data/DfD/R assumption or current forecast	High
Start of Production	See above/costs for late changes	Forecast	Cost tracking (see above)	Engineering actions	Engineering actions (insurance classes)/ Insurance forecast	Forecasting tools	Based on tear-down data/current EOL forecast	Medium
1st Market Launch	Market data	Current Legislation	Market data (but limited to that year)	See above	See above	Forecasting tools	See above	Medium
Full Market Penetration	Market data	Current Legislation	Market data	Market data	Market data	Market data available after several years	See above	Low

(continued)

Table 2.5 (continued)

Cost Info (right) being in LC stage (down)	Product price	Taxation	Operation cost (energy)	Maintenance/ warranty costs	Insurance costs	Residual value	End-of-Life (EoL) cost/revenues	Uncertainty
First Products Disposed	Market data	Current Legislation	Market data	Market data	Market data	Market data	Based on first invoices	Lowest
Uncertainty	Lowest	Low	Medium	Medium	Medium	Medium	Highest	

market cost are relevant for conventional LCC. Alternative cost and social cost are the primary cost types for societal LCC. For environmental LCC, a choice must be made. This is because the system suggests an alternative type of costs that includes the net transfer payments from and to governments. However, market cost may be adequate in practice if the analysis is focused on businesses and consumers. Alternative costs and social costs are extremely difficult to specify, so the choice for market prices could be justified as the only practical choice [56].

The second level reflects the entirety of the considered system. In principle, all stages in the life cycle should be included. However, from the point of view of an individual organization, the sum of its internal costs is equal to its cradle-to-gate cost. This cost can be obtained by combining its value with the costs of external purchases of goods and services, including waste management services. Nevertheless, this does not correspond to the full system costs. The third level categorizes the life cycle stages in more detail, and may be especially useful to track overheads, which are often neglected in LCA systems specification. The activity types distinguished in Table 2.6 may be easily expanded on systematically by using the EU-nomenclature as developed in NACE and its US equivalent NAICS. Both involve several hundred well described activity types. These classifications of activity types have a global origin. They are based on the ISIC classification of the UN. In the fourth level, the most specific cost elements are distinguished. Table 2.6 only shows some examples of cost elements. The cost elements chosen vary case-by-case and depend on conventions or on the level of detail aspired [56].

Table 2.6 Overview of cost categories [56]

Level	Cost category		
1st level: In economics	Budget cost, market cost, alternative cost, social cost		
2nd level: Life cycle stages	Knowledge development (incl. R&D), primary production (materials, food, energy), components production, manufacturing, use, end-of-life management		
3rd level: Activity types	Development, extraction, purchase, sales, reuse, management		
	Design, agricultural production, schooling, public relations, recycling, administration		
	Research, testing, packaging, transport, maintenance, waste processing, infrastructure		
4th level: Exemplary cost elements	Management: Material cost, Energy cost, Personnel cost, Machinery cost, Transport cost, Disposal cost, Revenues, End of Life value Supplementary: Service cost, Tooling cost, Storage cost, Taxes, Warranty, Assurances, Infrastructure cost, Building/asset costs, Settlement cost, Control cost, Financing cost, Appliance cost	Direct/indirect taxes excises, levies subsidies	Damage prevention, waste water treatment exhaust gas reduction rehabilitation costs

Table 2.7 Limitations of methods for monetarization of externalities

Method to estimate externalities	Issues	Resulting cost level
Willingness to pay Willingness to accept	• Knowledge of interviewees about actual impacts of externalities • Dependency on economic situation of interviewee and regional context • Lack of reward for interviewee to establish a thorough answer • Strategic answers • Answers only valid for the time of survey • Answers influenced by the wording of questions • (Too) difficult ethical questions (value of life)	Highest costs
Differences in market prices	• Differences are rarely the impact of only the considered externality • Small data set, lack of representativeness • Only for ideal market (no subsidies, full transparency of all market players)	Theoretically a good match for the current market evaluation of externalities
Repair, prevention costs	• Depends on opportunity to repair/prevent • Remaining damage or additional benefit from repair/prevention (e.g. new windows improve noise and energy)	Lowest costs if there is remaining damage
Insurance costs, insurance premium	• Rarely just due to the considered externality • Only for ideal market (no monopolies/oligopolies, full knowledge of market players)	In theory good match for current market evaluation of externality

(continued)

Table 2.7 (continued)

Method to estimate externalities	Issues	Resulting cost level
Certificate price	• Few aspects covered only (mainly climate aspects; no social externalities) • Only emission not impact related • Only for ideal certificate market	In theory good match for current market evaluation of externality
All methods	• Future generations not included • Anthropocentric estimates; ethical limits • Results valid for limited region and time span but many externalities are beyond these limits • Cost dependent on method	

©Wulf-Peter Schmidt 2024. All Rights Reserved

The second step of cost aggregation for some LCC studies (see also Table 2.3) is the discounting step. This step requires good justification of whether or not a discounting of future costs is carried out and what discounting rate is chosen.

Discounting represents the time preference of when costs and benefits occur. It accounts for the time value of money (i.e. 1€ or 1$ in 10 years does not have the same value as 1€ or 1$ today) and discounting converts all costs and profits over the life cycle time into their present value. Discounting also accounts for future risk factors.

> LCC costs are categorized in economic terms, by life cycle phases, by activity types, or cost elements. Discounting of future costs conveys the present value, future risks & opportunities.

For example, discounting balances the difference between the value-return on an investment in the future and the money invested in the present. In this context, the so-called Net Present Value (NPV) considers whether the difference between the present value of cash inflows and the present value of cash outflows over time is positive. "NPV is used to measure the costs and benefits, and ultimately the profitability, of a prospective investment (initial investment C_0) over time (T). The discounting rate (r) takes inflation and returns into account and features particularly in capital budgeting and investment planning" [64]:

$$\mathrm{NPV} = -C_0 + \frac{C_1}{1+r} + \frac{C_2}{(1+r)^2} + \cdots + \frac{C_T}{(1+r)^T} \tag{2.24}$$

If the NPV calculated by this formula is positive, the investment makes economic sense. The discounting rate of investments should not only consider the interest rate but also adjust for risks and opportunity costs.

The discounting rate itself depends on the purpose and context of the LCC. If a conventional LCC is carried out from the perspective of a company or consumer, the discounting rate is based on capital interest rates, inflation, opportunity costs, required return-on invest, and/or general profit rates. It is usually between 5 and 15%. A societal LCC usually has rather low discounting rates –0.1% for resource depletion, 0.01% for climate change and 0.001% for toxicity are suggested values. However, their justification is difficult, as it must consider how long impacts will stay relevant [54, 56].

For conventional LCCs, discounting rates are derived from market data or internal financial benchmarks. For societal LCCs, the discounting rates are lower and justification is difficult.

Figure 2.10 shows the impact of different discounting rates on the impact of a future cost or profit element on the final LCC result. The value of such future costs/profits is halved at a discount rate of 1% roughly every 70 years, at a discount rate of 0.1% every 700 years, etc. The discounted future costs or profits in the year T (C_T) is calculated as follows:

$$C_{D,T} = \frac{C_T}{(1+r)^T} \tag{2.25}$$

For example, if a company for plastic bottles is investing 1 million € or $ in energy saving measures to save each year 100,000 € or $ the savings will never compensate for investment at a 10% discounting rate. At a discounting rate of 1% it will take more than 10 years that the savings justify the investment—what is too long for most companies.

2.3.4 Interpretation

The interpretation of LCC results assesses the hotspots and differences between the available options. The interpretation must consider the uncertainties of data (see for example Table 2.5). If external costs are covered by the LCC, the limits and systematical biases of the data source for externalities are important (see Table 2.7).

It is recommended to conduct a sensitivity, Monte-Carlo and/or scenario analysis, and look at different scenarios before coming to a conclusion. In a review of 133 LCC studies 48% included uncertainty analysis [54].

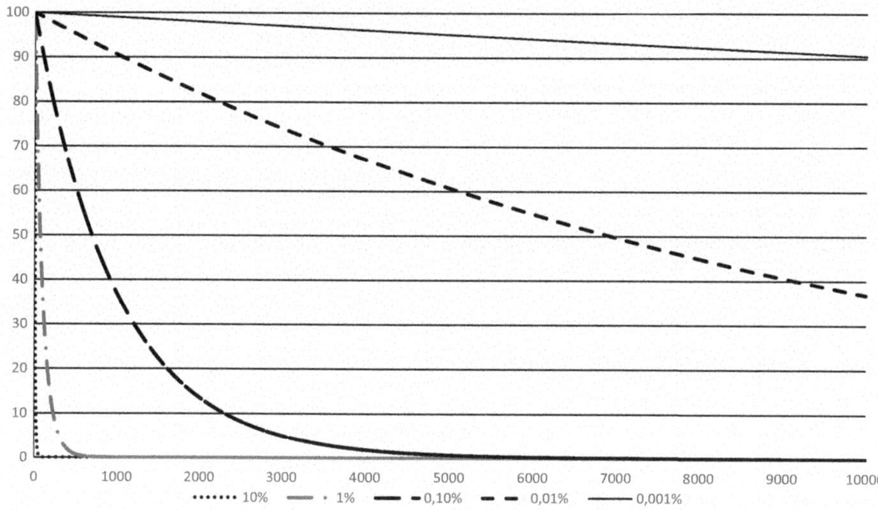

Fig. 2.10 Effect of different discount rates on importance of a future cost/profits. ©*Wulf-Peter Schmidt 2024. All Rights Reserved*

In the conclusion phase, it is not recommended to simply add up all environmental, social, and economic costs and profits/values in one figure, because of the bias and subjectivity involved in monetarizing external costs. For example, the Value Balancing Alliance methodology [65] suggests an accounting approach for upstream, downstream, and own operations where all these environmental, social, and economic impacts (positive and negative) are monetarized. However, they should only demonstrate the materiality of impacts in the different dimensions. Concluding on the positive or negative net value of an organization to society by adding up all negative external and internal costs and positive values [66] can lead to a situation where profits compensate for/justify/hide human rights costs. This is ethically unjustifiable and inappropriate—also having in mind the above-mentioned concerns about the monetarization of externalities.

2.4 Social Life Cycle Assessment (S-LCA)

Social Life Cycle Assessment (S-LCA) studies vary widely in their methodology which leads to inconsistencies and an even lower comparability of results as for environmental LCAs. A literature review has dated the origin of S-LCAs back to a work by Fava in a SETAC publication in 1993 where social welfare is introduced as an impact category for an environmental LCA [67]. The review concludes that for a long time the "scientific community has neither agreed upon a standardized methodology nor on a set of indicators…" [21]. Nevertheless, standardization work has led to ISO/CD 14075 which lays down "Principles and framework for social life cycle assessment". The standardization of S-LCAs by UNEP/SETAC guidelines remains as close to the environmental Life Cycle Assessment as possible (see Sect. 2.2) [15, 68]. The structures of the assessments are very similar even though LCAs focus on environmental flow impacts while S-LCAs consider social and organization-related aspects along the life cycle as well. This similarity is reflected in the origins of S-LCAs (see above) and work for a combined Social and Environmental LCA (SELCA) [69] as "an analytical tool for profiling and evaluating the interaction between the social and technological systems within the life cycle of given service" (Table 2.8).

2.4.1 Goal and Scope Definition

It is important to state the purpose and the audience of the S-LCA study. This allows the chosen approach and focus to be understood. Analogous to other LCAs, the goal could be to identify hot spots, support a sustainable product or process or organizational design, assess, check, and/or report on potential performance and improvements—in an S-LCA from a social perspective.

The definition of a quantifiable functional unit as a reference value for the results (see also Sect. 2.2.1) is especially difficult for S-LCAs. This is because they cover both physical and social flows and phenomena for which a causal relationship with the functional unit can be debated. For example, a technically well-defined functional unit like "moving 1 ton of food in a 1 m^3 container by 100 km at minus 20 °C" is difficult to link to social impacts like gender equality, hunger, or other UN social Sustainable Development Goals. In addition to the technical utility, social utility is also relevant. Social utility is defined as "a range of social aspects such as time requirement, convenience, prestige etc." [70]. For example, the product utility "providing a daily-basis vehicle to the local community for one year" goes beyond the functional unit of "Riding a bike for 1 km each day for one year in a certain country". This widens the perspective of the S-LCA to different stakeholder groups like local communities and local employment opportunities.

> Functional units of S-LCAs need to reflect the technical & social utility; the system boundaries need to include all socially significant effects that may extend beyond the physical flows.

The scope in terms of system boundaries and type of data must be defined in such a way that the goal of the study can be reached. The system boundaries should be set from a physical perspective (technological processes, economic flows along the value chain) and an effect perspective (relations among life cycle stakeholders as a result of the activities carried out) [68]. The cut-off criteria need to be justified, systematically applied, and validated. A cut-off criterion like "social significance" makes most sense for S-LCAs and needs to be validated in an iterative process. The

Table 2.8 Comparison of environmental and social life cycle assessment—see also [69]

Phase of the study	Characteristics
Goal and scope	• The product utility must be described in functional terms in both E-LCAs and S-LCAs. S-LCAs go further by also requiring that practitioners consider the social impacts of the product use phase and function • Product utility goes beyond the functional unit (or the function itself) and considers the perception of the consumer in regard to what the product provides also beyond its function including cultural and social values, desires and satisfaction from customer perspective. Product utility can be defined in technical (quality, functionality etc.) or in social terms (convenience, prestige, etc.) • While E-LCAs encourage the involvement of stakeholders (beyond the commissioners) during the peer review of the study, S-LCAs encourage such "external" stakeholders to be involved in providing input on impacts during the assessment itself • In S-LCAs, the absence of a subcategory in the study must be justified. In S-LCAs, these subcategories are classified both by "stakeholder categories" and by "impact categories". In E-LCAs, subcategories are classified only by "impacts categories" • While both, environmental and social impacts, are sensitive to location, E-LCA LCIA methods are typically not site-specific. S-LCAs may require site-specific LCIAs in some cases and may also depend on information about "political" attributes, such as the country and its laws
Life Cycle Inventory	• Data for "activity variables" is collected and used more often in S-LCAs than in E-LCAs (e.g. number of working hours used as an estimate for the share of each unit process in the product system). In E-LCAs such data is used when data about impacts is not available or if these variables are the focus of the investigation (see for example Fig. 2.3) • In contrast to E-LCAs, S-LCAs can and do make use of data that is reported/gathered also at the enterprise level, such as labor practices. S-LCAs will always attempt to gather information that is related to the facility where the unit process is located. However, for many subcategories the information is available solely at the enterprise/organization level • Subjective data is sometimes the most appropriate data type to use in S-LCAs. Abandoning subjective data in favor of more "objective" data introduces greater uncertainty in the results, not less • The balance between quantitative, qualitative, and semi-quantitative data will generally be different with more quantitative data for E-LCAs and more qualitative data for S-LCAs • The data sources will differ between S-LCAs and E-LCAs (stakeholders, databases, etc.). The data collection steps and methods will also vary (e.g. mass balances are not applicable for S-LCA)
Life Cycle Impact	• The characterization models are different • The use of performance reference points (e.g. thresholds) is specific to S-LCA • S-LCAs encompass both positive and negative impacts of the product life cycle, beneficial impacts in E-LCAs occur seldomly
Interpretation	• The significant issues will differ • Information on the level of engagement of stakeholders is included in S-LCAs

©Wulf-Peter Schmidt 2024. All Rights Reserved

cut-off criterion "available resources" runs the risk of not covering all important aspects and missing the goal of the study (Fig. 2.11).

An alternative to this process-oriented S-LCA study with functional units and system boundaries analogous to an environmental LCA is Social Organizational LCAs (SO-LCA) [68]. These look at the social impacts of an organization, considering the value chain for all its products and services (see also Sect. 2.6). Such an approach is especially advantageous for social impacts that cannot easily be allocated to individual processes/products in a linear fashion. Examples include the social impacts of corporate policies. It is more appropriate to replace the functional unit with activity variables that cover the entire organization and are not limited to individual products of an organization.

The following goal and scope aspects must also be specifically included for S-LCAs [68]:

- Stakeholders to be included.
- Types of impacts, impact categories, and subcategories to be considered.
- Inventory indicators and data related to those impacts.
- Methods for impact assessment.
- The activity variable to be used (to determine the relative importance of each unit process). Note: this item is not required for SO-LCAs. "Activity variables" are a measure of an activity/process that are related to the activity/process output and can be scaled by the activity.

Typical stakeholder categories include workers, local communities, value chain actors (e.g. suppliers), consumers, children, and society. Alternative categories are possible if justified. Relevant stakeholders are identified from the characteristics of the impact, their representativeness, legitimacy, and completeness [68].

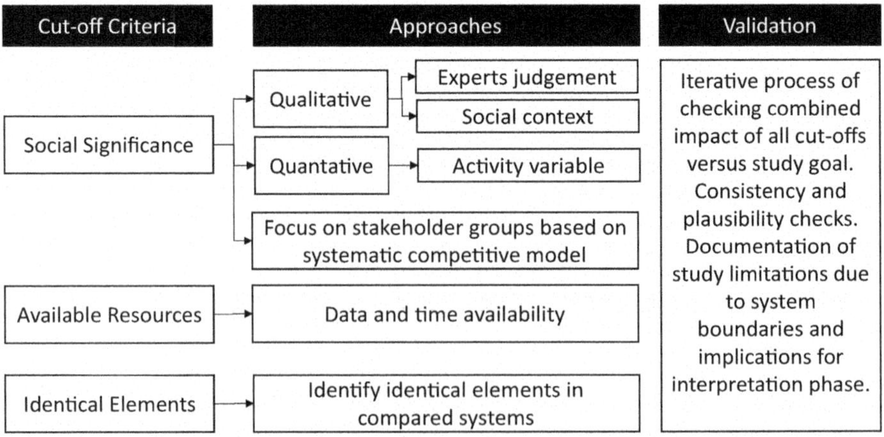

Fig. 2.11 Cut-off criteria in S-LCAs. ©*Wulf-Peter Schmidt 2024. All Rights Reserved*

Workers, local communities, value chain actors like suppliers, consumers, children & society are potential stakeholder categories of S-LCAs that should be covered by impact categories.

Ideally, external and internal stakeholders directly provide input. The type of data (site specific data, corporate data, average industry data, etc.) must be defined considering practicalities as well as the purpose of the study.

The impact categories and their subcategories ideally include all relevant social aspects, preferably considering the Sustainable Development Goals and their targets as well. Methods can follow a "midpoint" (e.g. wage level) or "endpoint" perspective (e.g. standard of living)—similar to environmental LCAs.

2.4.2 *Inventory Analysis*

During the inventory analysis, the life cycle system is modeled according to the goal and scope definition and data is collected—usually virtually as site visits/audits are not often possible. Data needs to be validated and where possible related to the unit process and functional unit. This validation can be done by statistically relevant spot checks. As part of this process, the system boundaries could be refined (e.g. due to allocation approaches for co-products or end-of-life options, see also Sect. 2.2.2.2). A sensitivity analysis is conducted to check whether a system boundary change could lead to a significant change in the result. For quantitative data a 1 or 5% threshold is often suggested. For qualitative data, a check must be conducted to determine whether the system boundaries chosen exclude significant social impacts or social impacts that are not covered by other impact categories. Finally, the data is aggregated along the life cycle.

The number of worker hours for each process step can serve as an "activity variable" to determine a ratio for allocating different social data to the unit process and functional unit. It supports the determination of the importance of different processes to allow prioritization of data collection and/or data validation. It also identifies potential hot spots from a social perspective. When using worker hours as an "activity variable", it is important to determine all working hours, including the unpaid, informal, and/or not documented work.

Data collection must be aligned with the impact assessment to ensure data is captured in such a way that allows later reference scales to be applied to non-quantitative data or enables a translation of data to the impact pathway assessment (see Sect. 2.4.3).

Data collected from databases, literature, surveys, interviews, audits, etc. can be allocated to the unit process/functional unit by activity variables like work hours (incl. informal/unpaid).

Actual data collection and validation can be done in several ways:

- Literature reviews and internet search—to identify data and relevant social items as well as for prioritization of data collection.
- Social Databases like the International Labor Organization (ILO) database, Social Hotspot database SHDB (www.socialhotspot.org), Product Social Impact Life Cycle Assessment (PSILCA; https://nexus.openlca.org/database/PSILCA), Fair Wage Database developed for performing S-LCIA with a focus on wage (www.see.tu-berlin.de), Maplecroft and others.
- Macro data like input–output data (Input–Output databases such as GTAP (Global Trade Analysis Project), EORA, EXIOBASE, WIOD (World Input Output Database)), and also national statistical agencies or intergovernmental organizations such as the International Labor Organization or the World Bank, or national/international sectoral organizations.
- Enterprise documentation (payrolls, etc.) and reports.
- Inquiries based on the consumer information act, environmental information freedom act, due diligence regulations etc.
- Governmental and NGO documentation (official statistics, complaints, etc.).
- Questionnaires and surveys.
- Interviews of stakeholders (workers, union representatives, citizens in neighborhoods).
- Audits (third-party, own, announced/unexpected).
- Anonymous hot line offerings.
- Focus groups, transect walks (see note below) and other participative methods.

Transect walks are a participatory exercise where stakeholders like members of the community, consumers, value chain actors physically walk through different areas of the neighborhood and worker's boarding houses, interviewing passers-by and drawing a map with observations of characteristics, risks/opportunities, and existing solutions after the walk.

Data collected for the different data indicators must be checked from the perspective of the different stakeholders, i.e. the social and socio-economic data needs to be triangulated. Data quality needs to be assessed and documented, considering reliability, completeness, and representativeness (see Table 2.9). Beyond this, it is very important to reflect on potential data ranges for later sensitivity analysis.

In the inventory phase background data for the later assessment is also collected. For example, the country-specific wages/cost of living are required to assess whether specific wages comply to the human right to adequate standard of living (article 25 of human rights) in the following impact assessment.

Table 2.9 Pedigree matrix for evaluating the data quality in S-LCA [68]

Indicator	Data quality scores				
	1	2	3	4	5
Reliability of source(s)	Statistical study or verified data from primary data collection from several sources	Verified data from primary data collection from one single source or non-verified data from primary sources, or data from recognized secondary sources	Non-verified data based partially on assumptions or data from non-recognized sources	Qualified estimate (e.g. by an expert)	Non-qualified estimate or unknown origin
Completeness conformance	Complete data for country-specific sector/ country	Representative selection of country-specific sector/country	Non-re-presentative selection, low bias	Non-re-presentative selection, unknown bias	Single data point/ completeness unknown
Temporal representativeness	<1 year difference to the time period of the dataset	<2 years difference to the time period of the dataset	<3 years difference to the time period of the dataset	<5 years difference to the time period of the dataset	Unknown or >5 years difference to the time period of the dataset
Geographical representativeness	Data from same geography (country)	Country with similar conditions or average of countries with slightly different conditions	Average of countries with different conditions or geography included, with large share, or country with slightly different conditions	Average of countries with different conditions, geography under study included, with small share, or not included	Data from unknown or distinctly different regions
Technical representativeness	Data from same technology and sector	Data from similar sector, e.g. within the same sector hierarchy, or average of sectors with similar technology	Data from slightly different sector, or average of different sectors with the sector under study making up a large share	Average of different sectors, sector under study included, with small share, or not included	Data with unknown technology/sector or from distinctly different sector

2.4.3 *Impact Assessment*

Similar to Life Cycle Impact Assessment (Sect. 2.2.2.2) where the elementary flows are "translated" into impacts to the environment, the social life cycle impact translates the physical and social flows into social impacts. These are defined by UNEP/SETAC [15] as "consequences of positive or negative pressures on social endpoints (i.e. well-being of stakeholders)" and as "consequences of social relations (interactions) weaved in the context of an activity (production, consumption or disposal) and/or engendered by it and/or by preventive or reinforcing actions taken by stakeholders (ex. enforcing safety measures in a facility)". Positive social impacts considered fall into the following three types [68]:

- Positive social performances that go beyond business as usual or grow the organization's own "social handprint" [71]—examples include corporate community involvement programs that support local communities.
- Positive social impact through presence (i.e. product or company existence)—examples are providing employment, infrastructure, etc. that improve several SDGs.
- Positive social impact through product utility—examples are water treatment plants that improve well-being of people (the functional unit usually reflects its product utility).

In this step, the relevant impact categories with the appropriate methods and models are first selected in line with the goal and scope definition (see Sect. 2.4.1). In the following step of classifying impact assessments, the inventory data is linked to the different sub-categories and impact categories. Finally, the sub-categories and impact categories are determined/calculated in the characterization step.

Similar to environmental LCAs, S-LCA impact categories can look at end-points (e.g. human well-being) or midpoints (e.g. safety; compare Sect. 2.2.3). However, a lot of research is still needed to establish satisfactory methodologies with robust, quantifiable cause-effect chains for all required impact categories. Therefore, S-LCA studies aggregate the Life Cycle Inventory indicators or sub-category results for the entire life cycle, i.e. relating these numbers to an "activity variable" like worker hours. This could result in statements like "99% of the workers' hours are proven to be free of slavery". Sub-categories can be introduced due to the heterogeneity of impact categories like human rights (based on 30 different articles). So far, they cannot usually be aggregated scientifically.

> Negative and positive social impacts exist—the latter due to product utility, by presence or by better performance. End-points (e.g. well-being) or midpoint (e.g. safety) methods exist.

This leads to several weighting and aggregation steps between inventory data and final impact categories relevant for the different stakeholders (see Fig. 2.12).

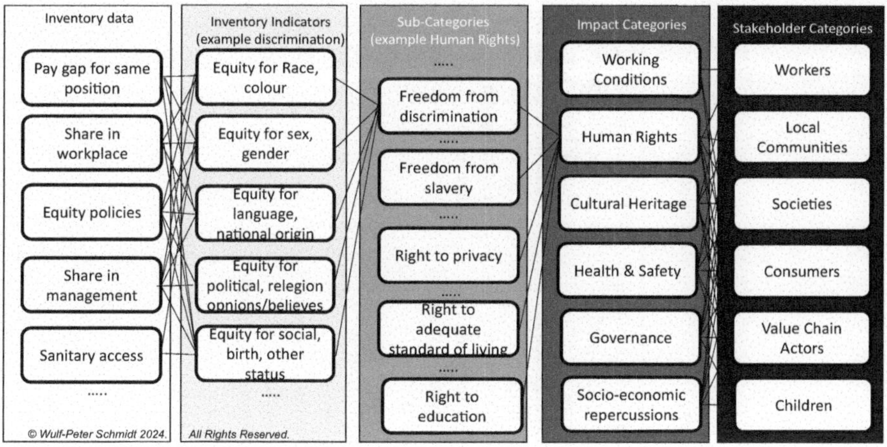

Fig. 2.12 Example for flow from inventory to impact and stakeholder categories

Looking at this from the other way, several impact categories are usually relevant for each stakeholder category as outlined in Table 2.10.

For transformational changes, consequential S-LCAs need to consider other types of social aspects (see also Sect. 2.2.1.2) to check for social implications and acceptance of the required change. An example of social impact categories is shown for those categories that are linked to the circular economy in Table 2.11.

In impact assessments, it is important to differentiate between potential social impacts determined from proxy inventory data that are calculated based on a cause-effect chain and actual social impacts that are based on observed and verified specific primary data (Table 2.9).

Different to environmental LCAs, impact category models may not always represent direct multiplication factors applied to quantitative elementary flows from life cycle inventory. Instead, they might be based on scorings or other systems that also translate qualitative information. An example is a reference scale with 5 different levels as suggested in the revised guidelines of UNEP/SETAC for social performance evaluation regarding social risk evaluation [68]:

+2	For ideal performance (e.g. wages significantly above a living wage), so low social risk
+1	For progress beyond compliance (e.g. wages above legal minimum and further improvements), so medium risk
0	For legal compliance and/or alignment to international standards (e.g. legal minimum wage paid)
−1	For non-compliant situations with actions for improvement taken (e.g. sometimes wages are below legal minimum because of deductions), so high social risk
−2	Non-compliance and no actions taken or no data (wages systematically below legal minimum), so very high social risk

Table 2.10 Impact subcategories per stakeholder category [68]

Worker	Local community	Value chain actor (excl. customers)	Customer	Society	Children
– Freedom of association and collective bargaining – Child labor – Fair salary – Working hours – Forced labor – Equal opportunities/ discrimination – Health & safety – Social benefits/social security – Employment relationship – Sexual harassment – Small-holders incl. farmers	– Access to material resources – Access to immaterial resources – Delocalization and migration – Cultural heritage – Safe and healthy living conditions – Respect of indigenous rights – Community engagement – Local employment – Secure living Condition	– Fair competition – Promoting social responsibility – Supplier relationships – Respect of intellectual property rights – Wealth distribution	– Health and safety – Feedback mechanism – Consumer privacy – Transparency – End-of-life responsibility	– Public commitments to sustain-ability issues – Contribution to economic development – Prevention & mitigation of armed conflicts – Technology development – Corruption – Ethical treatment of animals – Poverty alleviation	– Education provided in the local community – Health issues for children as consumers – Children concerns regarding marketing practices

Table 2.11 Examples for social impact categories of circular economy (CE) identified by [72] (modified)

Inventory indicators	Sub-category	Impact category	Stakeholder category
Structural changes in job market due to CE, closure of loops in sector	Local employment	Job creation/decline	Local communities
Attitude and knowledge barriers, not-in-my-backyard (NIMBY), level of participation, local impacts, community equity, etc	Community acceptance	Social acceptance	Local communities
Circular inter-firm networks, human rights violations, Supply and demand networks	Access to material/ natural resources,	Supply chain/resource risk/resilience	Value chain actors, society
Pollution levels, new transport routes	Workers' health	Health & Safety	Workers

Modification ©Wulf-Peter Schmidt 2024. All Rights Reserved

Reference scale S-LCAs or Type I S-LCAs typically assess the "social performance of activities of organizations in the product system based on specific reference points of expected activity (called performance reference points—PRPs). Reference Scale Assessments … focus primarily on the activities of companies in the product system and commonly consider their immediate evaluation (e.g. at inventory indicator), i.e. no further propagation of effects. As such, Reference Scale approaches do not commonly …establish a link between the activity and longer-term impacts. Rather, based on available information, they estimate the likely magnitude and significance of potential social impacts further down the line" [68].

Impact pathway S-LCIA or Type II S-LCAs "assess potential or actual social impacts by using causal or correlation/regression-based directional relationships between the product system/organizations' activities and the resulting potential social impacts …("characterization"). Here, the analysis focuses on identifying and tracking the consequences of activities possibly to longer-term implications along an impact pathway" [68]. Type II S-LCIAs correspond most with environmental LCIAs even though qualitative impact pathways are possible for this type of S-LCIA as well.

Quantitative or score-based impact categories look at performance relative to reference points (Type I) or relative to long-term implications along the impact pathway (Type II).

An example for a reference point approach is the following example of fair wages where the impact of wages can be compared directly to identified fair wages [68]:

The determination of fair wages depends mainly on three country/region-specific and/or product-specific parameters [68]:

(1) living wages—MLW_n is the minimum living wage by country, region, or sector [€/month], which must be paid to the worker to enable an adequate living standard for an individual and/or family in the country or region where process n is performed;
(2) working time—CWT_n is the contracted working time [% of month] for workers performing process n (for percentage calculation the vacation days are considered),
(3) income (in-)equality—IEF_n is the (squared) inequality factor [expressed in percentages] of the organization, region, or country where process n is performed.

To develop a characterization model, characterization factors for these parameters must be operationalized:

$$CF_{FW,n} = \frac{1}{MLW_n} \times CWT_n \times \left(1 - IEF_n^2\right) \tag{2.26}$$

where $CF_{FW,n}$ is the fair wage-related characterization factor [month/€] for process n representing the country, region, or sector specific conditions (Characterization factors translate the inventory results into the common unit of the category indicator).

The impact indicator "fair wage equivalents" (given in "fair wage potentials"—FWP_n) can then be calculated as follows:

$$FWP_n = \frac{RW_n}{RWT_n} \times CF_{FW,n} \tag{2.27}$$

where RW_n is the real (average) wage [€/month calculated over one year] that is paid to the worker employed in process n. RWT_n is the real working time [hours/week] of workers performing process n (including vacation days and unpaid overtime).

"Results are given as the distance of the wage from a fair wage ($FWP = 1$). If

(a) RW_n is smaller than the MLW_n, the resulting FWP_n will be <1, implying the greater the distance from the (minimum) targeted state, the lower the FWP_n;
(b) the real working time is equal to the CWT then no effect on the FWP_n occurs
(c) RWT_n is greater than CWT_n what means overtime work, the resulting FWP_n will be <1 and will keep getting smaller the more overtime the worker does. Hence, the characterization factors determine how far reality is from the minimum targeted situation" [68].

The following example illustrates how long-term impacts along the longer impact pathway can be quantified for child labor that causes lost income due to lost education.

Child labor can be assessed by the number of cases upstream of and within own operations. Each case can be counted as one year of missed education and each year of lost education leads to an average loss of income of 10% due to missed opportunities resulting from the lack of education. This can then be used in the Health Utility of Income (HUI) method. The HUI represents the contribution of income to the well-being of an individual at a country level (l for location) and is applied to convert the income change into the change of an individual's quality of life [63]:

$$\text{HUI}_l = \frac{\text{Change in Well-being}_l}{\text{Income change}_l} \tag{2.28}$$

Based on this, the disability-adjusted life years (DALY) can be calculated by multiplying it with the net present value (NPV) of the lost income of the children.

$$\text{Lost DALYs} = \text{HUI}_l \times \text{NPV of income lost} \tag{2.29}$$

Table 2.12 provides an overview on differences between these two types of S-LCIAs.

Ideally, S-LCA impact assessments cover all social issues of the UN Sustainable Development Goals [73]. SETAC/UNEP attempts this. However, this approach is rarely carried out in S-LCAs [21] and the allocation suggested by SETAC/UNEP has been criticized as unclear (Fig. 2.13).

2.4.4 Interpretation

The last step of S-LCAs is very similar to environmental LCAs and consists of

- An identification of hot spots of the analyzed products and services as well as any significant issues with the study in terms of study limitation due to system boundaries, data, and methodology.
- The evaluation of the study (completeness and consistency, sensitivity/Monte Carlo, and other analysis).
- Conclusions, recommendations, and reporting.

  S-LCAs in addition
- Report on the level of engagement with stakeholders.

Before this, a completeness check regarding the goal and scope definition is required along with data quality and consistency checks. This should be an iterative process to allow identified gaps to be closed. Then, scenario and sensitivity analysis should be conducted to examine the impact of the identified weak points of the S-LCA study on the potential ranges of results. A materiality analysis is recommended by UNEP/SETAC, where a "social matter (information, data, performance, impact, stakeholder) is material if it is of such significance and importance that it could substantially influence the conclusions of the study, and the decisions

Table 2.12 Difference between Type I and Type II Social Life Cycle Impact Assessments [68]

Characteristics	Reference Scale Approach (Type I)	Impact Pathway Approach (Type II)
Aim for users	Assessment of the product system in terms of its social performance or social risk	Assessment of consequences resulting from the product system, with a focus on potential social impacts
Relation between activity and potential social impacts	Relation assumed. Analysis focuses on activity and its direct evaluation, more specifically on its position in regards to performance reference points	Cause-effect relations established through (1) developed and justified characterization models or (2) qualitative descriptions of existing relations along an impact pathway
Scope of analysis	Variable coverage of life cycle stages. Variable coverage of multiple stakeholder categories and impact subcategories	Variable coverage of life cycle stages. Variable coverage of stakeholder and impact categories and their defined impact pathway (e.g. human health)—some studies focus on one impact pathway, others on many
Type of results	Social performance or risks, organized by impact subcategories, stakeholder categories, and/or life cycle steps usually considers social well-being as the area of protection	Impact categories comparable to E-LCIA toward defined Areas of Protection, such as well-being, addressing one or more stakeholder categories and/or life cycle steps
Nature of indicators	Qualitative, quantitative, or semi-quantitative with most studies focusing on the latter	Qualitative or quantitative with most studies focusing on the latter
Object of the assessment	Immediate activities and their effects	Usually short- and longer-term social consequences characterization
Existing S-LCA databases	Databases that focus on but are not limited to performance evaluation at risk level (e.g. SHDB & PSILCA)	Databases that focus on but are not limited to single or multiple impact pathway issues (e.g. fair wage, EXIOBASE extension)

©Wulf-Peter Schmidt 2024. All Rights Reserved

and actions based on those conclusions. Materiality is thus independent from the level of influence that an organization plays on the different phases of the product system under study" [68]. *Materiality* is a term often used in accounting but also in other areas and refers to identifying the issues that matter most to an organization, a stakeholder, a decision, etc.

Fig. 2.13 The S-LCA impact subcategories linked to the 17 SDGs [68]

> Interpretation of S-LCAs is similar to environmental LCAs but should focus even more on stakeholder engagement and uncertainty.

Like environmental LCAs, a critical review will increase the external credibility of the study.

2.4.5 Outlook

S-LCA research has to focus especially on the following challenges [21]:

- Linking LCA structures to social phenomena. Linking social impacts to a well-defined functional unit for example is difficult due to the weak causal relationship between the functional unit that is specific to a certain product or service and the broader social aspects and implications.
- Unclear, unexplained, or fragmented models and theories underlying social impact categories. The impact pathways can often not be quantified along the full pathway and cause-effect impacts can often not be isolated.

- Process versus organizational S-LCA, i.e. some impacts can be linked via activity variables (like working hours) that have a linear relationship between produced product and social impact while most social impacts are linked to the general organizational behavior of organizations/corporations instead.
- Missing clear link between the 17 Sustainable Development Goals (SDGs) with their 169 targets and 230 indicators and S-LCA impact sub-categories in most studies, partially explained by the difficulty to break down the global/national macro-level SDGs to the local S-LCA social impacts common to these studies (micro corporate level).
- Limited integration of positive impacts; in the past, 40% of studies did not consider positive impacts at all and the majority of those considering also positive impacts do not follow a consistence guideline.
- Generic versus site-specific data and the dependency on stakeholder participation, i.e. the data quality is very variable, just as it is for environmental LCAs. For S-LCAs, however, the situation is even worse, as social data may also depend on e.g. cultural, ethical, or national context that impacts the perception of how appropriate certain justice, equality, or ethical issues are. An example is child labor that is defined at different age levels (and may be seen by some cultures as supportive of the development and learning of children in some instances).
- Neglected stakeholder categories and the variety of impact sub-categories; children in particular are often not included as a stakeholder category. The impact sub-categories recommended by UNEP/SETAC (see Table 2.10) are often not considered consistently.
- Lack of comparability and transparency of S-LCA; in fact, most S-LCA studies are adjusted to individual cases without detailed explanation and rationale, undermining comparability, transparency, and acceptance of S-LCA studies.

2.5 Sustainable Life Cycle Management

Sustainable Life Cycle Management (S-LCM) makes life cycle thinking and product sustainability workable for businesses that aim to continuously improve and minimize their environmental and socio-economic impacts while also maximizing economic and social values along their value chains. For non-business organizations such as governmental organizations, S-LCM operationalizes life cycle thinking and sustainability by enabling the integration of policies and cooperations.

2.5.1 Shared Life Cycle Responsibility

A fundamental idea of sustainable life cycle management is the concept of shared responsibility between all life cycle stakeholders. Throughout the product life cycle, different stakeholders have the opportunity to improve social, economic, and environmental aspects in their spheres of influence. At the same time, all stakeholders have up-stream and down-stream partners. The resulting interfaces can provide incentives for life cycle partners to improve their performance. Part of this idea is to look for the most efficient solutions along the life cycle by organizing partnerships, information flows, and material loops.

Table 2.13 lists the roles of the life cycle stakeholders in improving the environmental, social, and economic performance of products and services. Sustainable life cycle management can be initiated by every one of the life cycle stakeholders. While big organizations may have certain advantages in organizing life cycle management (purchasing power, available resources, fewer dependencies on other stakeholders), small organizations can also leverage opportunities (lower complexity, agile processes and departments, faster and better internal communication, easier decision structure).

Table 2.13 Role of life cycle stakeholder in improving the environmental, social, and economic performance of products and services—based on [74]

	Role of industry (incl. suppliers and service)	Role of direct/ end consumer	Role of disposal/ recovery business	Role of policy, investors, insurances, etc.
Up-stream in the life cycle	Sustainable Supply Chain Management (social & environmental minimum standards)	Purchase more sustainable products & services while accepting premium	Inform end-user	Provide finances, rules that support sustainable supply, products, and infrastructure; penalize or tax all others
Own life cycle stage	Sustainable product and service design, environmental management, social standards, corporate citizenship	Sustainably use/ consume, follow instructions, minimize consumption of energy & materials	Establish sustainable recovery routes by generating competitive products	Stop subsidies, financing, policies for unsustainable behaviors & products Ensure own operations are a role model for sustainability
Down-stream in the life cycle	Provide product information & training, sustainable dealer standards	Direct products and materials to the appropriate collection, disposal, or recovery facilities	Information to producers, sustainable supply of high-quality recovery products	Provide finances, incentives, rules that support sustainable supply and products; penalize or tax all others

©*Wulf-Peter Schmidt 2024. All Rights Reserved*

Note simplified table—other, indirect stakeholders (for example research institutes, NGOs, media) also have roles in supporting sustainability (for example by providing sustainable technologies, information and education)

All life cycle stakeholders have opportunities to improve sustainability in their own, direct sphere of influence. In addition, they influence up- and down-stream stakeholders.

Numerous examples exist for each of the entries in Table 2.13. For the example of cars this includes the development and mass production of sustainable propulsion technologies, offering the necessary infrastructure (e.g. charging infrastructure), safe eco-driving, and the generation of recycled materials. However, the efficiency of these measures is quite different (Table 2.14).

The consumer is the key stakeholder as consumer demand for products and services is triggering in the end all environmental, social, and economic impacts. Changing consumer behavior toward more sustainable patterns and practices can be aided by policy, insurers, media, and producers to a certain extent (see Chap. 7). In addition, team-work between different stakeholders is crucial. This can be best demonstrated by looking at electric vehicles that need the support of the vehicle producer (offering these vehicles), charging network provider (ensuring widespread availability), electricity provider (ensuring sustainable electricity sources), and consumers (accepting premiums and being open to different technology and ranges, applying eco-driving). Such an integrated approach is the best way to efficiently and effectively improve the key impacts of vehicles.

The example of passenger vehicles also illustrates the importance of governments in shaping specifically consumer purchasing and driving behavior in a more sustainable way. The most effective method for achieving change is making non-sustainable products unattractive. This is shown by the most successful markets for Electric Vehicles—in particular Norway but also lately the Netherlands. In these countries, punitive taxation schemes demanding several tens of thousands of Euros for CO_2 emitting vehicles and clear end dates for the registration of new vehicles that are not zero-emission have been instrumental.

Table 2.14 Efficiency of exemplary actions of different life cycle stakeholder for passenger vehicles (percent reduction compared to base case)

	Propulsion option 1	Propulsion option 2	Eco-driving	40 kg recycled plastics[a]
Life Cycle Global Warming Potential	−15%	−55%	−25%	−0.03%
Summer Smog Potential	−20%	−80%	−30%	~0%
Life Cycle Cost	−1%	−15%	−5%	~0%
Social Impacts	Negligible	Livable, emission free cities, job losses	Improved safety	Negligible

©*Wulf-Peter Schmidt 2024. All Rights Reserved*
[a]Only theoretical option

2.5.2 *Life Cycle Sustainability Assessment*

Sustainable Life Cycle Management can be supported by the three basic Life Cycle tools (Life Cycle Assessment—see Sect. 2.2, Life Cycle Costing—see Sect. 2.3, Social Life Cycle Assessment—see Sect. 2.4) or ideally by a Life Cycle Sustainability Assessment (LCSA) combining the three Life Cycle tools into one integrated, consistent assessment [75]. The UN Life Cycle Initiative developed ten principles for harmonized LCSA practices [76]:

1. Indicate shortcomings in the cause-effect mechanisms connecting inventory results to mid- and endpoints of an area of protection.
2. Alignment—LCSA must consider the stages of goal and scope, inventory analysis, impact assessment, and interpretation, in accordance with ISO 14040:2006.
3. Completeness—in terms of the 3 sustainability dimensions and the life cycle.
4. Consider and include perspectives of all stakeholders.
5. Consider product utility co-benefits beyond the functional unit (e.g. for local communities).
6. Materiality of system boundaries i.e. all relevant and significant processes for any dimension are included.
7. Consistency of system boundaries, assumptions, data, models etc.
8. Transparency in all aspects of the LCSA.
9. Explicit communication of trade-offs.
10. Caution when compensating negative and positive impacts within a pillar or among the three pillars. At the very least present negative and positive impacts separately, and report weighting principles transparently if applied.

> LCSAs are more than just the sum of LCA, S-LCA and LCC. They require alignment, consistency and interactions between all methodological choices and conclusions.

An assessment of 193 LCSA studies [77] suggests that the ten principles are a feasible, solid basis from which LCSA practice can be enhanced. However, these ten principles are not yet fully implemented as a harmonized basis for LCSAs in practice. Challenges still exist with respect to understanding relationships between the three dimensions of sustainability covered and how to deal with trade-offs in decision-making. Issues identified in the review of 193 LCSA studies included [77]:

- Poorly understood overlap between environmental, social, and economic dimensions if for example human health issues are covered in LCAs with S-LCA.
- 21.8% of studies did not consider any stakeholder, 26.6% only one and 6.2% considered 5 or more stakeholders. Only 3.1% of studies considered children as stakeholders while more than 65% considered workers.

- Product utility was usually not considered.
- On average, 6 environmental impact categories, 5 social impact categories, and 3 economic impact categories were considered per LCSA study.
- No study provided sufficient information on the materiality of selected unit processes and the criteria behind this selection.
- Most studies were not consistent in applying the required methods systematically.
- 48% of studies were not transparent at all in terms of inventory data or models used.
- 60.4% of studies were explicit on trade-offs.

These issues suggest further research is still required.

2.5.3 *Circular Economy Principles*

Sustainability can be leveraged by life cycle thinking based on circular economy principles [78]:

- Narrow—use less materials and energy. This includes design approaches (Chap. 4), sustainable production (Chap. 6) as well as sustainable/lower consumption patterns (Sect. 7.1).
- Slow—use longer. This requires both suitable design for durability and repairability (Sect. 4.1) as well as consumption patterns (Sect. 7.3).
- Regenerate—substitute toxic or hazardous materials and processes (Sect. 4.6.2) by regenerative biomass resources.
- Cycle—use again. This includes all types of keeping materials and energies in the loop (Sect. 8.2)—also using partnerships across different life cycles.

All these steps require cultural changes, customer acceptance, and active support. There is a strong link between a circular economy and life cycle management in terms of shared life cycle responsibility (see Sect. 2.5.1). New business models can be established.

2.6 Sustainability Organizational Life Cycle Assessment

The environmental, social, and economic Organizational LCAs (SO-LCAs) compile all inputs, outputs, and potential sustainability impacts of the activities associated with an organization and its product and service portfolio along the organization's value chains. Different standards exist for the environmental, social, or economic dimension of an Organizational LCA individually including ISO 14072 [79], UNEP [68, 80] and EU [47]. Ideally, they are used together in SO-LCAs. An SO-LCA follows the same steps as process Life Cycle studies such as LCSAs, i.e. goal and scope definition, inventory, impact assessment/aggregation per cost category, and interpretation. The main differences to the process-oriented sustainability LCAs are the goal definition, system boundaries, and functional units (called *reporting organization*) i.e. the goal and scope definition step is very different.

2.6.1 Goal and Scope Definition

SO-LCA goals can include gathering insights into the own value chain from a sustainability perspective, the identification of performance, risks and opportunities, providing decision support, etc. However, UNEP does not recommend using O-LCAs for comparisons between organizations "due to the lack of a consistent basis for comparison between organizations" [68]. This is mainly because there are almost never organizations with the same product portfolio, vertical integration, etc. This recommendation is partially in conflict with common benchmarking practices where organizations in the same or even different sectors and segments are frequently compared with each other from a sustainability perspective as well (e.g. MSCI [81], Global 100 [82]). This is done by relating, for example, emissions to the revenue of the companies. However, such an approach systematically selects for higher priced products and services and does not correspond with a sustainability perspective that relates everything to human "needs" like the Brundtland sustainability definition does [83]. In any case, ISO 14072:2014 categorically denounces comparisons between organizations drawn from organizational LCAs if these publications will be disclosed to the public. The application of process-oriented LCAs, S-LCAs, and LCCs to organizations can have various goals including the identification of hot spots, tracking performance at a certain point in time or over a certain period of time and supporting decisions from a sustainability perspective.

In SO-LCAs' goal and scope definition, the studied organization is clearly defined including which products, operations, facilities, and sites of the organization are covered. The reference period (ideally 1 year or one season for seasonal business), *reporting flow*, system boundary, allocation procedures, impact assessment methodology and types of impacts, data, and data quality requirements are clarified. The type of interpretation to be used as well as any main assumptions, value choices, and optional elements and limitations, the type of critical review as well as the type and format of the report are also delineated.

"The reporting flow is a measure of the outputs of the reporting organization…and should answer the questions What? How much? How well? How long?" [80] This reporting flow can be measured in number of products, weight of output, revenue, hours of service, etc.

For SO-LCAs, the organization is the reference unit/reporting organization for the analysis in terms of its product portfolio i.e. the amount and nature of goods and services provided by the organization over the reporting interval [47]. This reporting organization needs to be clearly defined and can be the whole organization or—if transparently justifiable—parts of it (regions, brands, business divisions, facilities). The method of consolidating information is important in cases where Joint Ventures, subsidiaries, or similar structures are included. In such a case it must be indicated whether consolidation is done by equity share or by control method i.e. fully include a subsidiary if there is either financial or operational control by the studied organization or fully exclude if there is no control.

SO-LCAs are similar to process LCAs; the organization serves as a reference unit instead of a functional unit. Financial/operational control and the value chain define system boundaries.

The system boundaries must be justified and aligned with the goal of the study—as is done in product LCAs, product S-LCAs, or product LCCs. Usually, the entire up- and downstream value chain from cradle-to-grave must be covered. Depending on the goal of the study, additional aspects might be covered like business travel, employee commuting, working-environment related activities (canteens, child-care facilities, etc.), capital equipment, leased assets, and in selected cases even customer transportation.

2.6.2 Other Phases of SO-LCA

Data collection can be done bottom up, top down, or as a mixture/hybrid of both approaches. A Bottom-up approach basically sums up individual Life Cycle studies of all the different products (or product clusters) of the reporting organization weighted by their share for the reporting period and supplemented by supporting activities. The top-down approach takes the organization as a whole and adds upstream (cradle-to-gate) models for all inputs and downstream models for all outputs of the organization. Data quality and types of sources will be different for the top-down approach and need to be considered in the interpretation.

Impact assessments of or aggregation of costs by cost categories as well as interpretation follow similar principles to the product LCA, S-LCA, and LCC.

References

1. S.H. Farjana, M.A.P. Mahmud, N. Huda, *Life Cycle Assessment for Sustainable Mining* (Elsevier, 2021)
2. E. Jacob-Lopes, L.Q. Zepka, M.C. Deprá, *Sustainability Metrics and Indicators of Environmental Impact—Industrial and Agricultural Life Cycle Assessment* (Elsevier, 2021)
3. E.G. Hertwich, Increased Carbon Footprint of Materials Production Driven by Rise in Investments. Nat. Geosci. **14**, 151–155 (March 2021)
4. D. Paraskevas, K. Kellens, A. Van de Voorde, W. Dewulf, J. Doflou, Environmental Impact Analysis of Primary Aluminium Production at Country Level. Procedia CIRP **40**, 209–213 (December 2016)
5. N. Hill, S. Amaral, S. Morgan-Price, T. Nokes, J. Bates, H. H. Helms, K. Biemann, N. Abdalla, J. Jöhrens, E. Cotton, L. German, A. Harris, S. Haye, C. Sim, Determining the Environmental Impacts of Conventional and Alternatively Fuelled Vehicles Through LCA, 2020
6. ACEA, *Life Cycle Assessment in the Automotive Industry—Position Paper* (ACEA, Brussels, 2021)
7. T.L. Karl, Oil-Led Development: Social, Political and Economic Consequences, 2007
8. H. Ritchie, You Want to Reduce the Carbon Footprint of Your Food? Focus on What You Eat, Not Whether Your Food Is Local, Our World in Data, 2020. [Online]. Available: https://ourworldindata.org/food-choice-vs-eating-local. Accessed 25 Nov 2023
9. J. Fleming, *The Biofuels Myth: Why 'Sustainable Aviation Fuels' Won't Power Climate-Safe Air Travel* (Center for Biological Diversity, Tucson, Arizona, US, 2022)
10. Boundary Layer Technologies, ARGO, Boundary Layer Technologies [Online]. Available: https://www.boundarylayer.tech/argo. Accessed 5 Mar 2023
11. ISO14040, *Environmental Management—Life Cycle Assessment—Principles and Framework,* 2006
12. ISO (14044:2006), Environmental Management—Life Cycle Assessment—Requirements and Guidelines/Amd 2:2020, 2020
13. ASTM, American Society for Testing and Materials, Standard Practice for Measuring Life-Cycle Costs of Buildings and Building Systems E917, 1995
14. S.K. Fuller, S.R. Petersen, *Life Cycle Costing Manual: For the Federal Energy Management Program; NIST Handbook 135* (U.S. Government Printing Office, Washington, 1996)
15. UNEP, SETAC Life Cycle Initiative, *Guidelines for Social Life Cycle Assessment of Products* (UNEP, 2009)
16. M. Möller, F. Antony, R. Grießhammer, J. Gröger, R. Prieß, *PROSA—Product Sustainability Assessment. Leitfaden PROSAplus* (Öko-Institut e.V. Institut für angewandte Ökologie, Freiburg, 2021)
17. W.-P. Schmidt, F. Butt, Life Cycle Tools within Ford of Europe's Product Sustainability Index. Case Study Ford S-MAX & Ford Galaxy. Int. J. LCA **11**(5), 315–322 (2006)
18. EU Commission, Commission Recommendation on the Use of the Environmental Footprint Methods to Measure and Communicate the Life Cycle Environmental Performance of Products and Organisations, vol. C(2021) 9332, Brussels: 2021
19. W. McDonough, M. Braungart, *Cradle to Cradle: Remaking the Way We Make Things* (North Point Press, New York, 2002)
20. A. Ciroth, D. Hunkeler, G. Huppes, K. Lichtenvort, G. Rebitzer, I. Rüdenauer, B. Steen, *Environmental Life Cycle Costing* (SETAC-CRC, Pensacola, FL, 2008)
21. L. Pollok, S. Spierling, H.-J. Endres, U. Grote, Social Life Cycle Assessments: A Review on Past Development, Advances and Methodological Challenges. Sustainability **13**(10286) (2021)
22. D. Hunkeler, K. Saur, G. Rebitzer, W.-P. Schmidt, A.A. Jensen, H. Strandorf, K. Christiansen, *Life Cycle Management* (SETAC Press, Pensacola, FL USA, 2004)
23. UNEP, SETAC; ed. by S. Valdivia, C.M. Lie Ugaya, G. Sonnemann, J. Hildenbrand, *Towards a Life Cycle Sustainability Assessment—Making Informed Choices on Products* (U. L. C. Initiative, Ed., UNEP, 2011)

24. European Commission, *Understanding Product Environmental Footprint and Organisation Environmental Footprint Methods* (Publications Office of the European Union, Brussels, 2021)
25. Korea Transportation Safety Authority, Ministry of Land, Infrastructure and Transport, Proposal of Level Concept of Vehicle Cycle for A-LCA Evaluation Approach. Proposal for the Informal UN-ECE Working Group Automotive LCA, Geneva, 2023
26. G. Rebitzer, T.F.R. Ekvall, D. Hunkeler, G. Norris, T. Rydberg, W.-P. Schmidt, S. Suh, B.P. Weidema, D.W. Pennington, Life Cycle Assessment—Part 1: Framework, Goal & Scope Definition, Inventory Analysis, and Applications. Environ. Int. **30**(5), 701–721 (2004)
27. A. de Beaufort-Langeveld, N. van den Berg, K. Christiansen, R. Haydock, M. ten Houten, S. Kotaji, E. Oerlemans, W.-P. Schmidt, H.K. Stranddorf, A. Weidenhaupt, P. White, *Simplifying LCA: Just a Cut?—Final Report of SETAC EUROPE LCA Screening and Streamlining Working Group* (SETAC, Brussels, 1997)
28. VDI, VDI 4600:2012 Cumulative Energy Demand (KEA)—Terms, Definitions, Methods of Calculation. VDI-Gesellschaft Energie und Umwelt, Düsseldorf
29. M. Ritthof, H. Rohn, C. Liedtke, *Calculating MIPS—Resource Productivity of Products and Services* (Institute for Climate, Environment and Energy, Wuppertal, 2002)
30. W.-P. Schmidt, G. Fleischer, High Recycling Rates = High Reduction of Environmental Burdens? in *R'97 Congress Proceedings, 4.-7. February 1997,* vol. II, pp. 137–142, 1997
31. W.-P. Schmidt, *Ökologische Grenzkosten der Kreislaufwirtschaft* (TU Berlin, Berlin, 1997)
32. T. Ekvall, A. Björklund, G. Sandin, K. Jelse, *Modeling Recycling in Life Cycle Assessment* (IVL Swedish Environmental Research Institute, Gothenburg, 2020)
33. Joint Research Center, Harmonised Rules for the Calculation of the Carbon Footprint of Electric Vehicle Batteries (CFB-EV), EU Commission, 2023
34. EU Commission, *Commission Recommendation 2021/2279 on the Use of the Environmental Footprint Methods to Measure and Communicate the Life Cycle Environmental Performance of Products and Organisations* (EU Commission, Brussels, 2021)
35. D.L. Schrijvers, P. Loubet, B.P. Weidema, To What Extent Is the Circular Footprint Formula of the Product Environmental Footprint Guide Consequential? J. Clean. Prod. **320** (2021)
36. G. Fleischer and W.-P. Schmidt, Functional Unit for Systems Using Natural Raw Materials. Int. J. Life Cycle Assess. **1**(1), 23–26 (1996)
37. S. Nilson, Y. Virtanen, *Environmental Impacts of Waste Paper Recycling* (International Institute for Applied System Analysis, Luxemburg, Austria, 1993)
38. SETAC, *Life Cycle Assessment Data Quality: A Conceptual Framework* (Wintergreen, Virginia, USA, 1992)
39. A. Edelen, W.W. Ingwersen, The Creation, Management, and Use of Data Quality Information for Life Cycle Assessment. Int. J. Life Cycle Assess. **23**(4), 759–772 (2018)
40. A. Ciroth, R. Arvidsson, Data Quality in Life Cycle Inventories, in *Life Cycle Inventory Analysis* (Springer, 2022)
41. B.P Weidema, C. Bauer, R. Hischier, C. Mutel, T. Nemecek, J. Reinhard, C.O. Vadenbo, Overview and Methodology. Data Quality Guideline for the Ecoinvent Database Version 3. Ecoinvent Report 1(v3), The Ecoinvent Centre, St. Gallen, 2013
42. JRC, *European Platform on LCA | EPLCA—ILCD International Life Cycle Data system.* European Commission [Online]. Available: https://eplca.jrc.ec.europa.eu/ilcd.html. Accessed 20 Nov 2023
43. Federal LCA Commons—A Central Point of Access to a Collection of Data Repositories for Use in Life Cycle Assessment [Online]. Available: https://www.lcacommons.gov/. Accessed 20 Nov 2023
44. G. Myhre, D. Shindell, F.-M. Bréon, W. Collins, J. Fuglestvedt, J. Huang, D. Koch, J.-F. Lamarque, D. Lee, B. Mendoza, T. Nakajima, A. Robock, G. Stephens, T. Takemura, H. Zhang, Anthropogenic and Natural Radiative Forcing, in *Climate Change 2013: The Physical Science Basis. Contribution of Working Group I to the Fifth Assessment Report of the Intergovernmental Panel on Climate Change* (Cambridge University Press, Cambridge, UK, 2013)
45. D.W. Pennington, J. Potting, F. Gunnar, L. Erwin, J. Olivier, T. Rydberg, G. Rebitzer, Life Cycle Assessment: Current Impact Assessment Practice (Part 2). Environ. Int. **30**(5), 721–739 (2004)

46. Leiden University, CML, CML-IA Characterisation Factors, 5 September 2016. [Online]. Available: https://www.universiteitleiden.nl/en/research/research-output/science/cml-ia-characterisation-factors
47. M. Damiani, N. Ferrara, F. Ardente, *Understanding Product Environmental Footprint and Organisation Environmental Footprint Methods* (Publications Office of the European Union, Luxembourg, 2022)
48. IPCC, The Intergovernmental Panel on Climate Change [Online]. Available: https://www.ipcc.ch/. Accessed 20 Nov 2023
49. M. Muhl, M. Berger, M. Finkbeiner, Distance-to-Target Weighting in LCA—A Matter of Perspective. Int. J. Life Cycle Assess. **26**, 114–126 (2021)
50. ISO 14044:2006: Environmental Management—Life Cycle Assessment—Requirements and Guidelines, 2006
51. W.-P. Schmidt, J. Sullivan, The Value Debate: Weighting of Life Cycle Assessments in a Global Context: Global Diversity Exists and Has to Be Valued. Int. J. LCA **7**(1), 250, 6–10 (2002)
52. V. Prado, M. Cinelli, S.F. Ter Haar, D. Ravikumar, R. Heijungs, Sensitivity to Weighting in Life Cycle Impact Assessment (LCIA). Int. J. Life Cycle Assess. **25**, 2393–2406 (2020)
53. A. Laurent, B.P. Weidema, J. Bare, X. Liao, D. Maia de Souza, M. Pizzol, S. Sala, H. Schreiber, N. Thonemann, F. Verones, Methodological Review and Detailed Guidance for the Life Cycle Interpretation Phase. J. Ind. Ecol. **24**(5), 935–1185 (2020)
54. S. L. Rodrigues, E.A. da Silva, Implementation of Environmental Life Cycle Costing: Procedures, Challenges, and Opportunities. Int. J. Life Cycle Assess 2024
55. M. Schmidt, Total Cost of Ownership TCO—How to Uncover the Hidden Costs of Owning, in *Business Encyclopedia* (Newbury St Boston, 2023)
56. D. Hunkeler, K. Lichtenvort, G. Rebitzer, *Environmental Life Cycle Costing* (SETAC, Pensacola, Florida, 2008)
57. A.C. Pigou, *The Economics of Welfare* (Macmillan, London, 1918)
58. W.-P. Schmidt, Life Cycle Costing as Part of Design for Environment—Environmental Business Cases. Int. J. Life Cycle Assess. **8**(3), 167–174 (2003)
59. A. Endres, Ökonomische Grundprobleme der Messung sozialer Kosten. List Forum **11**(4), 251–269 (1982)
60. L. Wicke, F. Schafhausen, W. Schulz, H.-D. Haasis, *Betriebliche Umweltökonomie – Eine praxisorientierte Einführung* (Franz Vahlen, München, 1992)
61. EMBER, Carbon Price Tracker—The Latest Data on EU ETS Carbon Prices [Online]. Available: https://ember-climate.org/data/data-tools/carbon-price-viewer/. Accessed 20 Nov 2023
62. OECD, Mortality Risk Valuation in Environment, Health and Transport Policies, Paris, 2012
63. Value Balancing Alliance, VBA METHODOLOGY V0.1—Impact Statement—Topic Specific Method Paper: Environment, Frankfurt a. M., Germany, 2021
64. Discount Rate Formula: Calculating Discount Rate [WACC/APV] Paddle [Online]. Available: https://www.paddle.com/resources/discount-rate-formula. Accessed 16 Nov 2023
65. Value Balancing Alliance, International Foundation for Valuing Impacts, General Methodology 1—Conceptual Framework for Impact Accounting (exposure draft), 2023
66. BASF, Value-to-Society—Quantification and Monetary Valuation of BASF's Impacts on Society, Ludwigshafen, 2017
67. J.A. Fava, *A Conceptual Framework for Life-Cycle Impact Assessment* (Society of Environmental Toxicology and Chemistry and SETAC Foundation for Environmental Education, Pensacola, FL, 1993)
68. C. Benoît Norris, M. Traverso, S. Neugebauer, E. Ekener, T. Schaubroeck, S. Russo Garrido, M. Berger, S. Valdivia, A. Lehmann, M. Finkbeiner, G. Arcese, *UNEP Guidelines for Social Life Cycle Assessment of Products and Organizations 2020* (United Nations Environment Programme (UNEP), 2020)
69. M. O'Brien, A. Doig, R. Clift, Social and Environmental Life Cycle Assessment (SELCA). Int. J. Life Cycle Assess. **1**(4), 231–237 (1996)

70. R. Grießhammer, C. Benoît, L.C. Dreyer, A. Flysjö, A. Manhart, B. Mazijn, A.L. Méthot, B. Weidema, *Feasibility Study: Integration of Social Aspects into LCA* (Öko-Institut, Freiburg, 2006)
71. R. Husgafvel, Exploring Social Sustainability Handprint—Part 1: Handprint and Life Cycle Thinking and Approaches. Sustainability **13**, 11286 (2021)
72. A. Luthin, M. Traverso, R.H. Crawford, Assessing the Social Life Cycle Impacts of Circular Economy. J. Clean. Prod. **386**, 135725 (2023)
73. United Nations, SDG, 25 September 2015. [Online]. Available: https://documents-dds-ny.un.org/doc/UNDOC/GEN/N15/291/89/PDF/N1529189.pdf?OpenElement. Accessed 27 Feb 2023
74. W.-P. Schmidt, Strategies for Environmentally Sustainable Products and Services. Corp. Environ. Strat. **8**(2), 118–125
75. W. Klöpffer, Life Cycle Sustainability Assessment of Products. Int. J. Life Cycle Assess. **13**, 89–95 (2008)
76. S. Valdivia, J.G. Backes, M.S.G. Traverso, S. Cucurachi, J.B. Guinée, T. Schaubroeck, M. Finkbeiner, N. Leroy-Parmentier, C. Ugaya, C. Peña, A. Zamagni, A. Inaba, Principles for the Application of Life Cycle Sustainability Assessment. Int. J. Life Cycle Assess. **26**, 1900–1905 (2021)
77. N. Leroy-Parmentier, S. Valdivia, P. Loubet, G. Sonnemann, Alignment of the Life Cycle Initiative's "Principles for the Application of Life Cycle Sustainability Assessment" with the LCSA Practice: A Case Study Review. Int. J. Cycle Assess. **28**, 704–740 (2023)
78. Circle Economy, *The Circularity Gap Report 2023* (Circle Economy, Amsterdam, 2023)
79. ISO 14072 Environmental Management—Life Cycle Assessment Requirements and Guidelines for Organizational Life Cycle Assessment, 2014
80. UNEP, Guidance on Organizational Life Cycle Assessment, Paris, 2015
81. Corporate Sustainability Insights," MSCI [Online]. Available: https://www.msci.com/corporate-sustainability. Accessed 2 Nov 2023
82. Corporate Knights, The 2024 Global 100: Overview of Corporate Knights Rating Methodology, 2023
83. United Nations, Our Common Future (The Brundtland Report): World Commission on Environment and Development, UN, 1987

Chapter 3
Sustainable Strategy Development and Planning

Abstract A fundamental starting point for fostering sustainability in organizations is to establish strategies and plans that address key sustainability challenges and opportunities. The integration of sustainability into the overall (business) strategy and plan for the wider organization is crucial. The deeper sustainability is ingrained into the organizational and business strategy and the more this integration is driven from the top of the organization, the better the outcome will be. Any sustainability strategy will fail if it is developed in isolation without executive buy-in or if it is added as an after-thought. In fact, such approaches often lead to supposedly large and well-publicized improvements, but lack substance and efficacy. This chapter describes how traditional business strategy development can be used for and enriched by sustainability.

Keywords Materiality assessment · SWOT analysis · Futuring · Benchmarking · Strategy · Design thinking · Change management · Transformation

A fundamental starting point of sustainability is establishing strategies and plans that address key sustainability challenges and opportunities. Most important is the integration of sustainability in the overall (business) strategy and plan for the general organization, avoiding an isolated strategy/plan. The more sustainability is integrated into the organizational and business strategy and the more this integration is driven from the very top of the organization the better. Any sustainability strategy will fail if developed in isolation without executive buy-in or if completed as an after-thought. In fact, such approaches often lead to superficially great sustainability reports and press releases, but lack substance and effectiveness.

Of course, a strategy driven from the top of the organization is not guaranteed to be successful. Strategies often fail due to unwilling and uncooperative middle management. In addition, the organizational culture must encourage the engagement of employees. In consequence, the development of a successful sustainability strategy must involve all relevant stakeholders within and outside the organization.

© The Author(s), under exclusive license to Springer Nature Switzerland AG 2024

W. Schmidt, *Solutions For Sustainability Challenges*,
https://doi.org/10.1007/978-3-031-63624-0_3

Sustainability strategies need to be an integral part of the overall organization's strategy. Plans must be supported by all levels rather than consisting of a stand-alone approach.

The first step to establishing a strategy is to gather external and internal input. Important input types in the development of the strategy include:

1. External and internal stakeholder and issue scan
2. Double materiality assessment that determines how sustainability issues impact an organization as well as how the organization impacts on the planet and its people
3. Benchmarking
4. Futuring and scenario planning
5. Do no significant harm criteria
6. SWOT analysis

All of these inputs for general strategies and plans offer an opportunity for Life Cycle and sustainability perspectives.

3.1 External and Internal Stakeholder and Issue Scan

A thorough scan of the external stakeholders, their perspectives, and external issues is an important first step. It is based on research and/or consultation of key stakeholders via interviews, workshops, surveys, and other means. A good approach is to look at the world and the own organization through the lens of these stakeholders and consider their issues.

Customers and clients are key stakeholders. They are the "immediate recipients or beneficiaries of the services or products". Understanding the customer is key to any successful strategy. Other potential stakeholders include owners/investors, employees, up- and downstream value chain representatives, Non-Governmental Organizations (NGOs), regulators, media, etc. Again, a broad life cycle perspective needs to be taken.

The analysis must be comprehensive by including at the very least political, economic, social, environmental, technological, resource/knowledge, legal, security/safety, regulatory, demographic, religious, and societal aspects. Analysis must consider what the current and future trends of these aspects are and how they may impact the organization and its sustainability strategy [1].

It is important to cover the impacts on and of the organization's strategy from and on the different life cycle stakeholders respectively. The interests of all life cycle stakeholders, their employees, investors, communities, and the environment need to be reflected.

Such an analysis will help anticipate any challenges and opportunities of the chosen strategy.

> The stakeholder and issue scan needs to look through the lens of all life cycle stakeholders, NGOs, regulators, media, etc. at the own organization.

Additionally, vulnerabilities need to be analyzed from a life cycle perspective. For example, human rights concerns along the supply chain, supply chain disruptions, and changes to the regulatory framework are important aspects of this step.

3.2 (Double) Materiality Assessment

The Global Reporting Initiative (GRI) [2] requires organizations to determine material topics relevant to an organization i.e. topics that are material/significant enough to "have a direct or indirect impact on an organization's ability to create, preserve or erode economic, environmental and social value for itself, its stakeholders and society at large". In other words, a Materiality Analysis or Assessment determines and prioritizes the most important sustainability impacts linked to the organization and its stakeholders. In order for this to happen, each item has to be considered through two lenses—what the potential positive or negative impact on the organization is and how important that impact is seen by its stakeholders. Issues include environmental, social, and governance (ESG) concerns and economic issues—in the entire life cycle. Governance issues address compliance topics especially in the own organization but also cover the supply chain and end-of-life operations or customer relations. The materiality assessment typically results in a portfolio showing the impact on businesses as well as the importance to stakeholders of these material topics (Fig. 3.1).

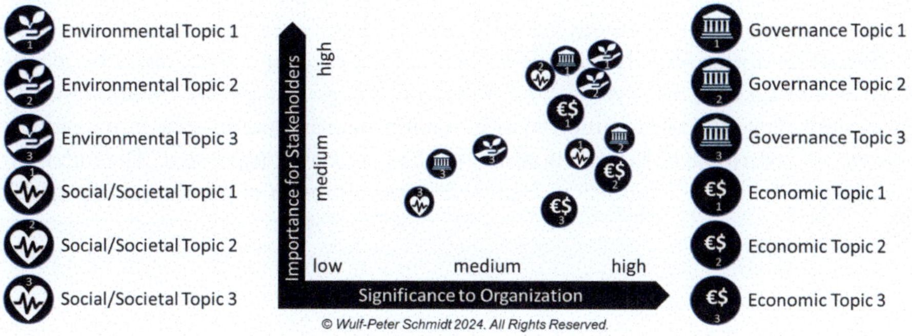

Fig. 3.1 Typical portfolio matrix representing material topics of an organization as part of a materiality assessment. © Wulf-Peter Schmidt 2024. All Rights Reserved

The portfolio is normally established by intense consultation of internal and external stakeholders in group or individual sessions, interviews, surveys, etc.

Materiality assessments rank potential positive and negative sustainability impacts from the perspective of stakeholders and the own organization.

Double materiality is "the union (in mathematical terms, i.e. union of two sets, not intersection) of impact materiality and financial materiality" [3]. It is meant to address the EU Corporate Sustainability Reporting Directive (CSRD) requirement for affected organizations to "include in the management report information necessary to understand the undertaking's impacts on sustainability matters, and information necessary to understand how sustainability matters affect the undertaking's development, performance and position."

The term *impact materiality* covers similar outcomes as the materiality assessment shown in Fig. 3.1. More precisely, it is defined as "a characteristic of a sustainability topic or information in relation to an undertaking" while "a sustainability topic or information is material from an impact perspective if the undertaking is connected to actual or potential significant impacts on people or the environment and is related to the sustainability topic over the short, medium or long term. This includes impacts directly caused or contributed to by the undertaking and impacts which are otherwise directly linked to the undertaking's upstream and downstream value chain" [3]. This definition originates from a life cycle perspective.

The term *financial materiality* is seen as separate from traditional financial reporting rules. In the context of sustainability, however, it is seen as "a characteristic of a sustainability topic or information in relation to an undertaking" while a "sustainability topic is material from a financial perspective if it triggers financial effects on undertakings, i.e. generates risks or opportunities that influence or are likely to influence the future cash flows and therefore the enterprise value of the undertaking in the short, medium or long term but are not captured by financial reporting at the reporting date. These risks and opportunities may derive from past events or future events and may have effects on future cash flows in relation (i) to assets and liabilities already recognized in financial reporting or that may be recognized as a result of future events or (ii) to factors of enterprise value creation that do not meet the accounting definition of assets (liabilities) and/or the related recognition criteria but contribute to the creation/maintenance of enterprise value. The latter are generally defined as « capitals» in frameworks promoting a multi-capital approach" [3].

Double materiality assessments rank potential positive or negative impacts on sustainability in terms of the impact the organization may have on them and vice versa.

The methodology for this step is still evolving, It offers great insights for strategy development as well. For example, if an organization (*undertaking*) is responsible for a significant share of global warming (e.g. in cement, steel, automotive, fossil energy industry), global warming will be a material impact that has the potential for future liabilities and risks to enterprise value. This increases the strategic need to change the business model which is reflected by the carbon neutrality strategies of most affected industries.

3.3 Benchmarking

Benchmarking is a continuous process that evaluates products, services, and practices against the strongest competitor or of the enterprise recognized as the leader [4]. Benchmarking is a multi-step process (number of steps vary between methods) including [5, 6]:

1. Plan/Self-analysis—without knowing the own processes and challenges it is difficult to determine what to learn from other organizations. The definition of target performance indicators for comparisons will be helpful during the later comparison. From a sustainability perspective, topics identified by the materiality analysis (Sect. 3.2) can be part of these indicators.
2. Research—this step defines the scope and focus of the benchmarking study, the organizations to be included as benchmarks as well as data already available.
3. Observe—collect complementary data and observe similarities and differences.
4. Analyze—all data is assessed and opportunities for change in the own organization are identified.
5. Adapt—the conclusions are adapted to the own organization and context.
6. Improve—in a continuous process of further refinement.

The Plan-Research-Observe-Analyze-Adopt-Improve (PROAAI) method can be implemented in different ways. The ISC Matrix is one suggestion for structuring and analyzing benchmarking data (Table 3.1).

Table 3.1 ISC matrix [6]—slightly adjusted

	Is	Should not be	Could be
What	What types of *(sustainability)* problems are tackled in this process/*action*?	What types of *(sustainability)* problems should not be tackled in this process/*action*?	What types of *(sustainability)* problem could be also tackled in this process/*action*?
Who	Who is a party in the process *within life cycle*? Who is responsible for the process/ *action*? Is external help used to set up the process?	Who should not be an actor in the process? Who should not be responsible for the process? Which other forms of help should not be used to set up the process?	Who might be a party in the process *within life cycle*? Who could be responsible for the process? Which other help could be used to set up the process?
When	How often is the process / *action* controlled? When is the process set up? When and how often is external assistance used?	How often should the process/*action* not be controlled? When is the process least powerful? When and how often should external assistance not be used?	How often could the process/*action* be controlled? When could the process be set up more effectively? When and how often is external assistance most relevant?
Where	In which departments has the process/*action* been implemented? In which interaction with stakeholders is the process most suitable?	In which departments should the process/*action* not be implemented? In which interaction with stakeholders should the process not be implemented?	In which departments could this process/*action* be used? In which interaction with stakeholders could this process be implemented?
How	How is the process/*action* seen by the Executive Committee/ *life cycle stakeholders*? How many meetings of the Executive Committee resulted in the process? How are human resources allocated to the deployment of this process, or, how many people are full-time for this process?	Should the process not be considered by the Executive Committee/*life cycle stakeholders*? How should human resources not be allocated to the deployment of this process, or, how many people should not be full-time for this process?	Could the process be considered by the Executive Committee/*life cycle stakeholders*? How could human resources be allocated to the deployment of this process, or, how many people could be employed full-time for this process?
Why	What are the results expected of the process/*action*?	What are the results you should not expect/*action*?	What are the results you could expect/*action*?

Adjustments © Wulf-Peter Schmidt 2024. All Rights Reserved

3.4 Futuring and Scenario Planning

To develop a resilient and sustainable strategy, it is important to be open-minded about the future and consider a wide range of potential scenarios of how the world may develop. This can be supported by futuring and scenario planning. "Futuring is an anticipatory decision-making process that leads to planning and program development ... it looks forward 10, 20, or 30 years, while planning looks ahead 1 to 4 years" [7].

Different anticipatory methods exist but it is recommended [7, 8] to adhere to the following sequence:

- Scanning and monitoring the external environment. In this context, "emerging issues" around regulatory and technological development, consumer and market trends, and other aspects are commonly considered. In the context of sustainability, it is also important to include environmental changes such as climate change/water stress, biodiversity changes as well as social trends, governance changes etc.
- Analyzing and challenging internal and external assumptions. This is an important step based on the many different types of disruptive events including financial crashes, pandemics, wars leading to energy and food crises as well as mass species extinction events and climate change.
- Creating multiple scenarios around emerging issue areas identified in the previous steps. The inclusion of *black swan events* [9] can also be helpful. Black swan events have major impacts but are surprising and do not conform to expectations. They are often rationalized in hindsight by suggesting that these events could have been predicted. Equally, *white swan events* that have major impacts but are statistically viable exist (like Covid-19: a global pandemic was expected with certainty but the specific disease and timing was unknown).
- Conducting issue vulnerability audits; this is meant to cover an "organization's vulnerability to change and disruption that may result from new competitors, new regulations, special-interest-group initiatives, scientific discoveries, ...media disclosure" but also includes disruptions arising from sustainability issues like climate change (i.e. climate resilience), water stress, human right concerns as well as technical innovations that change the market.
- Developing forecasts; such forecasts can encompass full descriptions of the future including confidence levels.
- Writing issue briefs; these concise documents should summarize all the intelligence gathered in such a way that is suitable for management.
- Identifying program champions, faculty, and staff who are ready to address predicted changes.
- Using the results of futuring to inform continuous improvement or a new strategy.

> Establishing different scenarios of potential futures must be open-minded, include major impact black (surprising) and white swan (statistically viable) events as well as disruptions.

3.5 Do no Significant Harm Criteria

A sustainable strategy does not "only" need to have a substantial, positive contribution for one or more sustainable development goal [10]. It also has to follow the precautionary principle of *do no significant harm*, so that none of these sustainable development goals is significantly harmed. There are even regulatory definitions for this, for example in the EU Taxonomy Regulation (Regulation (EU) 2020/852 [11]). Again, the whole life cycle of products and services must be considered, looking at climate change, the environmental status of water, the circular economy, any pollution, biodiversity, etc. Even the use of substances of very high concern is part of these regulatory considerations.

An example is electric vehicles that have a substantial, positive contribution for climate change (see also Fig. 2.2) but receive criticism for high water consumption in battery production or perceived limitations in the circular economy. While batteries have a wide range of circular economy opportunities (reuse/second life usage, various recycling technologies for Lithium, Cobalt, Nickel etc.) there is a high water demand that must be addressed, for example by water recycling. This is an example of how the strategic input of the *do no significant harm* criterion can support a strategic direction constructively by leading to mitigating proposals that address potential weaknesses and threats—rather than just blocking solutions that improve the present-day situation by excluding them due to a *do no significant harm* criterion. Such an approach risks resulting in the rejection of all viable options because everything may have disadvantages in some areas.

> The precautionary *do no significant harm* criteria need to be applied in a constructive way that identifies potential concerns while focusing on constructive mitigation opportunities.

A similar social taxonomy regulation may arise at a later point in time. However, for the development of a sustainable strategy it is important—beyond any regulation—to consider a *do no significant harm* approach to avoid unsustainable practices resulting from the chosen strategy that will undermine the viability and credibility of any sustainability strategy.

3.6 SWOT Analysis

The SWOT Analysis systematically evaluates Strengths, Weaknesses, Opportunities, and Threats. It was previously called SOFT, an acronym for safeguarding the Satisfactory; opening Opportunities; fixing Faults; or thwarting Threats [12]. Ideally, all/different internal and external life cycle stakeholders take part in establishing this analysis. This ensures a comprehensive perspective and also results in a participatory process that engages different stakeholders and increases the buy-in of the internal stakeholders at least. A SWOT analysis can be also used to assess different concepts and technological options (see example Table 4.6).

3.7 Establishing the Sustainability Strategy

If a strategy "is a set of choices about winning" [13], these choices need to be made with sustainability as a constructive and integral part of the choices that lead to winning in mind.

Based on the analysis of all the stakeholders, competitors, issues and their materiality, strengths-weaknesses-opportunities-threats, precautionary aspects as well as potential future scenarios, the main aspiration of a sustainability strategy can be determined. The aspiration should have a winning sustainability vision in mind.

The *North Star* is the one key part of an organization, the focal point of its whole strategy to which all actions of the organization are aligned with—the winning aspiration. It should be entrenched in the values of the organization and where applicable fully reflect the needs of its customers or target clients. It helps organizations—and individuals—to focus and ensure long-term alignment. It is similar to the categorical imperative in that it guides each (decision) situation towards the correct outcome.

Such a North Star could be formulated around the Paris Climate Accord commitments, biodiversity, human rights, and/or any other aspiration that has been identified as the consolidated outcome of all of the above-mentioned input. This north star should be articulated in an aspirational sentence that combines it with a potential for value creation of the organization that secures its long-term existence. For example, an organization could combine a commitment to climate targets with a production strategy vision like moving to 100% carbon neutral products by 2035. Going even further, companies could aim to contribute positively (rather than a net-zero approach) to carbon emissions by offering products or services that achieve a reduction in emissions (so-called scope 4 emissions). In healthcare, positive contributions to SDGs are often an integral part of companies' visions, an approach that could serve as a blueprint for strategies in the field of climate change.

> A sustainability strategy needs to have a convincing north star aspiration combined with a viable, sustainable value creation vision that secures the organization's long-term existence.

In terms of *technical sustainability management*, it makes sense to have a key component that is reflected in the objectives of each organization, department, manager, and employee. This could function as a base for performance reviews and renumeration.

Fig. 3.2 Strategy development based on strategic choices ([13] with modifications). *Additions* © Wulf-Peter Schmidt 2024. All Rights Reserved

Based on all the analysis and the initial winning aspiration, several strategic choices need to be made regarding where to play, how to win, what capabilities must be in place, and what management systems are required (see Fig. 3.2). These choices need to look at all life cycle stakeholders, implications, and opportunities.

3.8 Strategic Transformations

All of these theories are unfeasible when dealing with an organization that must transform fundamentally, as it is part of a sector without any sustainable future. Examples include suppliers of parts for automotive internal combustion engine in a region where automotive manufacturers and society/regulators decide to end this business practice. The following strategies can be observed in combination or as stand-alone strategies in practice:

1. Denial—continue business as usual, push for eFuels as solutions, shrink rather than change (Fig. 9.2).
2. Regional escape—continue business as usual but shrink (slow walk reaction to market)/stop doing business in the region and migrate to other regions where the old, unsustainable business is possible/allowed for a longer time, i.e. buying time for a real business transformation.
3. Diversifying—use the transitional period to maximize profits from the old business without investing in that part and use its profits to finance investments in new businesses.

4. Transforming—change the processes to be sustainable (e.g. conventional to green steel making) or innovate and supply parts for new businesses (e.g. for electrification) and find ways to finance the transition phase towards transformation (e.g. additional revenues like offering services or finding new investors).

Models for automotive manufacturers suggest that those who do not react until forced by regulation will likely be out of business in the long run. Those who try to follow the market and electrify in line with convenient market conditions (see above "regional escape") will stay profitable in the short-term but loose profitability and end up with significantly lower margins (EBIT) in the mid- and long-term due to the complexity of continuing to produce in the old, unsustainable as well as the new, sustainable way due to the low market share in the new market segment. In addition, they will exceed their CO_2 budget and shrink. Front-runners who are transforming quickly and determinedly towards full electrification will lose their profitability in the short-term but have higher margins than today in the long run and are the only type of automotive manufacturers in line with a well-below-2 °C pathway as set out by the Paris Climate accord. To finance the interim period of around 5 years, larger investments than profits and additional measures are needed (e.g. vehicle data monetarization, service offerings) and this period must be kept short by quickly reducing costs for batteries etc. [14]. Financial support from investors or governments might be needed.

Organizations in transformation need to act with urgency but require additional funding and income streams to bridge several years of high investments and losses.

To be successful in transformation, it is key to establish a "business model for sustainability, that is, a business model designed to maintain and create value in ecological, social and economic terms" [15]. Another key issue is to ensure that the entire organization, including employees, is fully on board with the transformation (see also Sect. 9.1.2).

Different organizations must find "patterns" for this transformation that best fit their culture and needs. These "patterns" are defined as "repeatable solution(s) to enduring environmental, social and economic challenges that arise when an organization aims to create, deliver or capture value in a sustainability-oriented manner". They fall into different groups, where each group addresses an overarching challenge organizations face in sustainable value creation like finding the right price, getting investments, closing the loop, etc. [15–17]. These groups consist of:

1 Service and performance patterns that support the transition from physical products to services. Examples include "pay for success" and "function as a service" (pay-per-use like mobility as a service or light as a service or pay-per-performance which is based on the achievement of pre-defined operational or financial outcomes, etc.).

2 Virtualize patterns—(dematerialized services) focuses on delivering products virtually rather than physically.
3 Pricing and revenue patterns to set prices and generate earnings from sustainable business offerings. Examples include differential pricing (aligned to different target groups), social Freemium with a free basic service and paid premium services as well as subscription models.
4 Financing patterns like crowdfunding, microfinance (small loans), profit reinvestment.
5 Product life extension and Eco-design patterns (see Chapter 4), like monetizing an extended product life by refurbishing for resale [18] or attracting new customer segments by offering products that can be upgraded over time (like smartphones with upgradable cameras).
6 Closing-the-loop and slowing material flow patterns by using by-product synergies, industrial symbiosis, online waste exchange platform, remanufacturing, repairing, reusing, take-back management, incentivized return schemes, upgrading, and circular economy ecosystems.
7 Supply chain patterns like green supply chain management, inclusive sourcing that prioritizes local, small or minority-owned businesses, micro distribution, and retail, virtual sales and distribution supported by augmented reality and AI, produce on demand, and short supply chains.
8 Industrial symbiosis patterns where material and energy flows are shared between organizations and former waste streams are repurposed to new resource flows.
9 Giving patterns like buy-one-give-one where donations are a fixed ratio to regular sales or a cost-free offering in return for data where the free offerings to a social group are financed by earning revenues from a second, commercial target group. Alternatively, capitalize on revitalizing ecosystems—like reforestation, reverting deserts—by marketing or by getting new resource sources as unique selling point.
10 Access provision patterns like Market Maker for neglected customer segments through tailored and inclusive approaches, e-transaction platforms, experience-based customer credits, last-mile grid services, value-for-money education/housing.
11 Social mission patterns empower social groups to become productive partners and transform their needs into demands like soup kitchen patterns, utilizing donor resources, socio-economic empowerment models with co-ownership for low-income entrepreneurs.
12 Cooperative patterns turn stakeholders into owners and decision makers.
13 Community platform patterns—like car-sharing. Such platforms are typically less profitable for manufacturers.
14 Encourage customer sufficiency or smarter consumption—like electricity providers who pay customers money if they consume less electricity (to save investment in new electricity generators) or use electricity at times of lower demand.

Sustainable business models can generate new income streams by offering new services, monetarize additional sustainability values and partner and empower life cycle stakeholders.

These patterns are not always perceived as beneficial, especially if not applicable, for instance during major industry transformation, or if further sustainable transformation solution categories such as technological change are required. Nevertheless, these patterns provide good examples.

To identify a sustainable business, classic tools like the Canvas tool can help (see Table 3.2).

Important for the success of sustainable business models is both the value proposition and how the value is captured i.e. turned into real income streams. Value capture can happen for example by [17]:

1 Premium prices for more sustainable products where the customer´s willingness-to-pay is enhanced by demonstrated savings for customers from reduced operating or disposal costs.
2 Directly reduced costs from lower input or emission control costs based on optimization in industrial symbiosis, lower consumption, shared asset costs, avoided investments etc.
3 Regular service and subscription fees (in addition or instead of once-off product sales), extended product warranties or take-back agreements.
4 Additional revenue from the sale of repaired or remanufactured products or from upgrading existing products.

3.9 Integrating Sustainability in Organizations—Sustainable Management

Organizing sustainability in organizations can be done in several different ways such as top-down or bottom-up, centralized or decentralized, and many more. The approach chosen should be based on the established integrated sustainability strategy and business model.

A top-down approach has the advantage of ensuring the necessary alignment of budgets and departmental objectives but requires the management to convince employees to fully embrace and support the needed change (see Sect. 9.1.2). A bottom-up approach usually takes longer to gain acceptance, lacks internal funding, and quickly comes into conflict with other priorities.

A centralized organization can ensure sufficient control and gather a critical mass of experts. However, it runs the risk of being isolated from the rest of the organization that simply does not follow directives. A decentralized approach in introducing sustainability could be employing a few central sustainability staff that support various sustainability teams in the main departments of the organization (like

Table 3.2 Canvas tool (adapted)

<table>
<tr>
<td rowspan="2">Key Life Cycle Partners
Who are key partners? Who are key internal players? Who are key suppliers? Which resources do we get from them? Which activities are they performing</td>
<td>Key Activities
What key activities does the sustainability opportunity require along the life cycle? What requires our value propositions? What our distribution channel? Revenue streams? Customer relationship? Relationship to other life cycle stakeholders & partnerships</td>
<td rowspan="2">Value Propositions
What value do we deliver to our customers and the different life cycle stakeholders? What problem of our customer/ Life Cycle partners do we solve? What customer and other life cycle stakeholders need to be satisfied?</td>
<td>Customer Relationships
What type of relationship does each customer segment expect us to establish and maintain? Which ones have we established?</td>
<td rowspan="2">Customer Segments
For whom are we creating value? Who are our most important customer segments?</td>
</tr>
<tr>
<td>Key Resources
For our value proposition, customer relationship, channels, partners</td>
<td>Channels
Through which channels do our customer segments want to be reached? How do we reach them today? Which works best? How are we integrating them in customer routines? What one is most cost-efficient?</td>
</tr>
<tr>
<td colspan="3">Cost Structure
Most important (Life Cycle) costs, of environmental/sustainable business model, what resources and activities are most expensive?</td>
<td colspan="2">Revenue Streams
For what value are customers really willing to pay? For what do they currently pay? How would they like to pay? Are there any new revenue streams possible along the life cycle?</td>
</tr>
</table>

For modifications © Wulf-Peter Schmidt 2024. All Rights Reserved

Purchasing, Manufacturing, Product Development). This approach ensures ownership and commitment of all main departments of an organization as long as these departments have objectives and Key Performance Indicators (KPIs) that reflect on their contribution to sustainability (see Fig. 4.7). The bigger the organization and the more independent the main departments are, the more important it is to focus on a decentralized implementation approach.

Sustainability can be organized top-down or bottom-up, with central or decentralized teams. Decentralized top-down approaches are most effective, especially for bigger organizations.

An example for a top-down organized sustainability strategy is recommended by ISO 14068:2023 [19] which details how a carbon neutrality strategy is best established and implemented. First, there needs to be a commitment from top management to ensure that all the necessary resources are in place and that the strategic direction of the organization is aligned with its carbon neutrality plan. This also means that the business model needs to be aligned. Otherwise, the clear hierarchy of ISO 14068:2023, which requires the plan to be based primarily on greenhouse gas reductions would not work. Carbon removals within the boundaries of the activity are only seen as the next step and offsetting is limited to the unavoidable small remainder. A carbon neutrality management plan needs to detail the timeline, system boundaries, baseline, targets, methodology as well as responsibilities, financial and human resources required and available.

In order to transform the own organization, agents of change and a controlled change management are required. Such management usually consists of the following steps:

1. Business opportunity—the opportunity to address the main challenges to sustainability must be linked to the organization's strategic priorities and objectives. With the help of *Design Thinking*, the business opportunity or issue is framed as a central question in a so-called *HMW* statement: *How Might We...*. This step raises awareness of employees.
2. Define the transformation—this step defines the business case for change, determines roles and responsibilities, and next steps needed to achieve the change. The biggest obstacles are also listed.
3. Diagnose the transformation challenge—gathering internal and external information. This can also be done with a Design Thinking mindset by immersing yourself in the context of your change, empathize with the people who will be impacted by the change, use observations of extremes and mainstreams as well as interviews as part of the *Gather Research & Inspiration* phase. In the *Identify Themes and Insights* phase, you look for the "a-ha" moments that reveal real human needs and opportunities for designing the transformation journey. Ideally, the desire for the transformation is increased in all stakeholders.
4. Design the transformation journey—Generating ideas and prototypes (experiment/small pilot) for the transformation can help make ideas tangible and allows learning through the pilots. The concept can be refined from these experiments. Knowledge about transformation methods is accumulated by business change agents.

5. Deliver the tested and refined transformation plan by communicating widely and integrating it into existing processes. By measuring the results, the sustained change must be demonstrated and if it is not, the process must again be refined or reinforced.

> Change management for sustainability needs to explain the need for transformation from the perspective of directly impacted stakeholders to ensure a constructive engagement.

Continued top-down sponsorship, clear ownership and accountability, an emphasis on communication with and engagement of those team members impacted by the change must be ensured. This can be done using interviews & focus groups as well as intensive socialization to make obstacles surface as early as possible and overcome resistance and build on the existing culture (Fig. 3.3).

Trust is another key component in change management: it is a requirement that allows internal and external stakeholders to embrace the requested change. Crucially, trust can only be earned over time by actions that are in line with previously stated and agreed upon (interim) targets or milestones.

When company leaders, neighbors, or other stakeholders need to be convinced to support a change, considering human behavior and convincing them by including stakeholders in deliberations remains important. Rather than starting by presenting

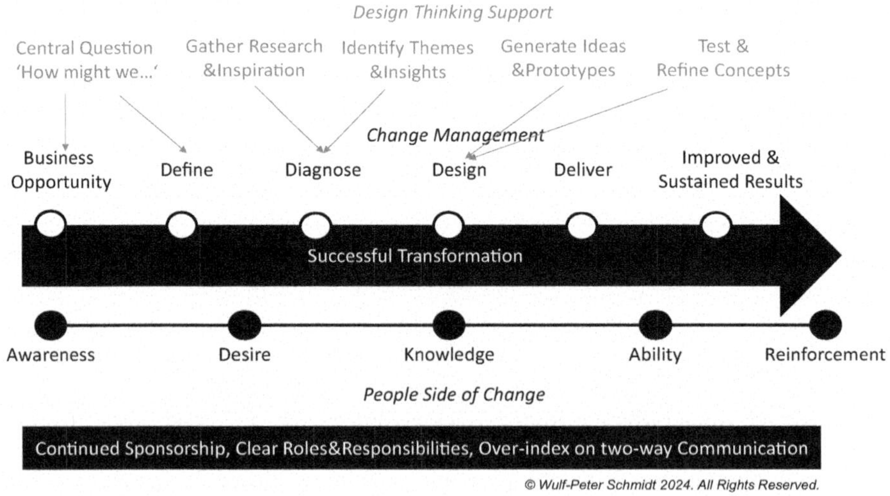

Fig. 3.3 Designing and delivering the transformation journey. © Wulf-Peter Schmidt 2024. All Rights Reserved

solutions, demands, or conclusions, successful change management starts by considering biggest concerns as well as personal or business objectives of each stakeholder. This helps them to come to the independent conclusion that sustainability is the best way forward to solve the issue they see as most pertinent (maybe without labeling the solution as "sustainable" if that is a contentious word). Innovation management needs to be part of the transformation especially when technologies are not ready for implementation yet.

References

1. United Nations, Strategic Planning Guide for Managers, New York
2. Global Reporting Initiative GRI, Consolidated Set of the GRI Standards, 2023
3. EFRAG, European Financial Reporting Advisory Group, Draft European Sustainability Reporting Guidelines 1—Double Materiality Conceptual Guidelines for Standard-Setting, Brussels, 2022
4. R.C. Camp, *Business Process Benchmarking: Finding and Implementing Best Practices* (ASQC Quality Press, Michigan, 1995)
5. J.-L. Maire, G. Büyüközkan, Methods and Tools for First Five Steps of Benchmarking Process, 1997
6. J.L. Maire, V. Bronet, M. Pillet, Benchmarking: Methods and Tools for SME. Benchmarking: An International Journal **15**(6), 765–781 (2008)
7. P.M. Sobrero, The Steps for Futuring. J. Exten. **42**(3), June 2004
8. W.C. Ashley, L. Morrison, Anticipatory Management: Tools for Better Decision Making. Futurist **31**(5), 47–50, September/October 1997
9. N.N. Taleb, The Black Swan: The Impact of the Highly Improbable, 2010
10. United Nations, SDG (25 September 2015) [Online]. https://documents-dds-ny.un.org/doc/UNDOC/GEN/N15/291/89/PDF/N1529189.pdf?OpenElement. Accessed 27 May 2023
11. European Union, Regulation (EU) 2020/852 of the European Parliament and of the Council of 18 June 2020 on the Establishment of a Framework to Facilitate Sustainable Investment, and Amending Regulation (EU) 2019/2088, European Union, Brussels, 2020
12. R.W. Puyt, L.F. Birger, C.P. Wilderom, The Origins of SWOT Analysis, *Long Range Planning*, 22 February 2023
13. A.G. Lafley, R.L. Martin, Playing to Win—How Strategy Really Works (Havard Business Review Press, Boston, 2013)
14. Deloitte, Pathway to Net-Zero—Mastering the Twofold Goal of Decarbonization and Profitability, 2023
15. F. Lüdeke-Freund, H. Breuer, L. Massa, Sustainable Business Model Design—45 Patterns, published by the authors, 2018
16. J. Binder, Unleashing the Power of Circular, in *12th Responsible Leadership Conference—Orientierung im Chaos, Die vielfältigen ESG-Anforderungen gemeinsam bewältigen, 22+23 November* (München, 2023)
17. T.D. Woldeyes, M. Muffatto, F. Ferrati, Archetypes of Business Models for Circular Economy: A Classification Approach and Value Perspective, in *International Conference on Sustainable Design and Manufacturing* (Singapore, 2023)
18. Renault, Re-factory, Paris, November 2020
19. ISO 14068-1:2023, Climate Change Management—Transition to Net Zero—Part 1: Carbon Neutrality, Vernier, Geneva, 2023

Chapter 4
Sustainable Product Development

Abstract Product development is commonly perceived as the most fundamental approach to reduce the environmental impact of a product over its entire life cycle. This life cycle includes the extraction of raw materials, material production, parts production, final assembly, use of product, and end-of-life operations including all transportation processes (aka cradle-to-grave or if material flows are closed in a circular economy approach also aka cradle-to-cradle). In contrast to purchasing and supply chain management (see Chap. 5), changes to product development have rather limited opportunities to improve social aspects beyond the product characteristics itself. Product development is an important factor in the economic success or failure of products and services. However, conventional product development approaches are limited in scope: economic life cycle impacts are not fully covered. Instead, the focus is on product cost or, in the best case, the cost of ownership of the first customer. In cases where repair costs are also considered, economic and environmental considerations overlap. Rarely, end-of-life costs are included in traditional product development. A selection of approaches for integrating sustainability into product development is described in this chapter.

Keywords Design for environment · Design for recycling · Design for dismantling · Design for sustainability · Product-service system · Sustainable materials · Lightweighting · Product sustainability index

Product development is perceived as a key starting point to influence the environmental impact of a product over its full life cycle—from extraction of raw materials, material production, parts production, final assembly, use of product, and end-of-life operations including all transportation (aka cradle-to-grave or if material flows are closed in a circular economy approach also aka cradle-to-cradle).

Compared to purchasing and supply chain management (see Chap. 5), product development changes have comparatively limited opportunities to improve social aspects beyond the product characteristics itself. Examples include designing a product so that disabled customers can easily use the product.

© The Author(s), under exclusive license to Springer Nature Switzerland AG 2024

W. Schmidt, *Solutions For Sustainability Challenges*,
https://doi.org/10.1007/978-3-031-63624-0_4

Product development is also an important factor in the economic aspects of products and services. However, conventional product development approaches are limited in terms of the economic aspects they cover. Often economic life cycle impacts are not fully covered. Instead, the focus is put on product cost or, in the best case, the cost of ownership of the first customer. In cases where repair costs are also considered, there is an overlap between economic and environmental considerations. Rarely, end-of-life costs are included in traditional product development.

A selection of approaches for integrating sustainability considerations into product development is described in this chapter.

4.1 Classical Design for X Approaches

There are different design strategies that include environmental aspects in product design (*Design for Environment*). *Design for X* is an approach that follows a certain design strategy that tries to optimize the design for the characteristics detailed in Table 4.1.

The emphasis in *Design for X* has changed over time. At first, *Design for Disassembly* was considered most important; however, the focus has since shifted via *Design for Recycling* and *Design for Environment*. Currently, *Design for Sustainability* is considered crucial. This development mirrored the broadening of focus in industry (and society).

Table 4.1 Design for X examples based on [1]

Life cycle stage	Design strategy
Raw Material	Design for resource conservation Design for low-impact materials (environmental, social)
Manufacturing	Design for manufacturing
Distribution	Design for efficient distribution (distance, mode, package)
Use	Design for energy efficiency Design for water conservation Design for minimal consumption Design for low-impact use Design for service and repair Design for durability
End-of-Life	Design for reuse Design for remanufacture Design for disassembly Design for recycling Design for safe disposal

For modifications © Wulf-Peter Schmidt 2024. All Rights Reserved

4.1.1 *Design for Disassembly and Recycling, as Well as Other More Recent Circularity Approaches*

From the 1970s to the beginning of the 1990s, a significant focus was placed on the issue of waste. Reuse and recycling of products were seen as a fundamental prerequisite of a sustainable approach for products given an anticipated shortage of resources [2]. Without closing the material loops, the danger of limiting growth was seen. Reuse and recycling were considered a target in themselves. To ensure reuse and recycling, product development was tasked with designing parts in complex products with this in mind. Washing machines, vehicles, etc. should have a design that allows easy dismantling of the targeted parts by addressing, for example, the accessibility of parts as well as the types and numbers of fasteners used. Design guidelines suggest using as few and as standardized fasteners as possible and making them easily accessible. Material selection should consider recycling aspects (e.g. material compatibility, reduced material diversity) while the use of recycled materials should increase demand for recycling products. Material compatibility tables (e.g. Fig. 4.1) convey the idea that non-compatible materials should not be part of the same product or component because they cannot easily be separated by common automatic processes (for example, because they have the same density or density ranges when considering their fillers). Such non-compatible materials cannot be recycled together without significant impacts on the quality of the recycled material.

Other Design for Recycling guidelines include minimizing material complexity by avoiding fiber reinforcements, additives, painting, coating, etc. [3].

> Design for Disassembly (DfD) and Recycling (DfR) rules include reducing the number of fasteners, lowering complexity, ensuring material compatibility, and using recycled materials.

Therefore, design constraints for a Design for Disassembly (DfD) and Design for Recycling (DfR) approach are wide-ranging and might conflict with other product attributes (craftsmanship, weight, packaging, etc.) and customer demands (e.g. soft-touch surfaces).

Doubts regarding such an approach increased. For example, in the automotive industry, the fundamental difference between vehicle assembly (workers trained to assemble a few parts for one or two vehicle models) and disassembly (mixture of hundreds of vehicle types, aged and damaged vehicles or screws, etc.) raised concerns. Real-world dismantling tests showed that only a small portion of the dismantling time can be addressed by-product design. Most theoretical links between design aspects and dismantling time such as length of screws, accessibility, visibility, etc. are overruled in practice by operators experience and work organization [4, 5]—or simply because the product is not dismantled but crushed in a shredder. This means that Design for Disassembly actions are often ineffective. This is especially

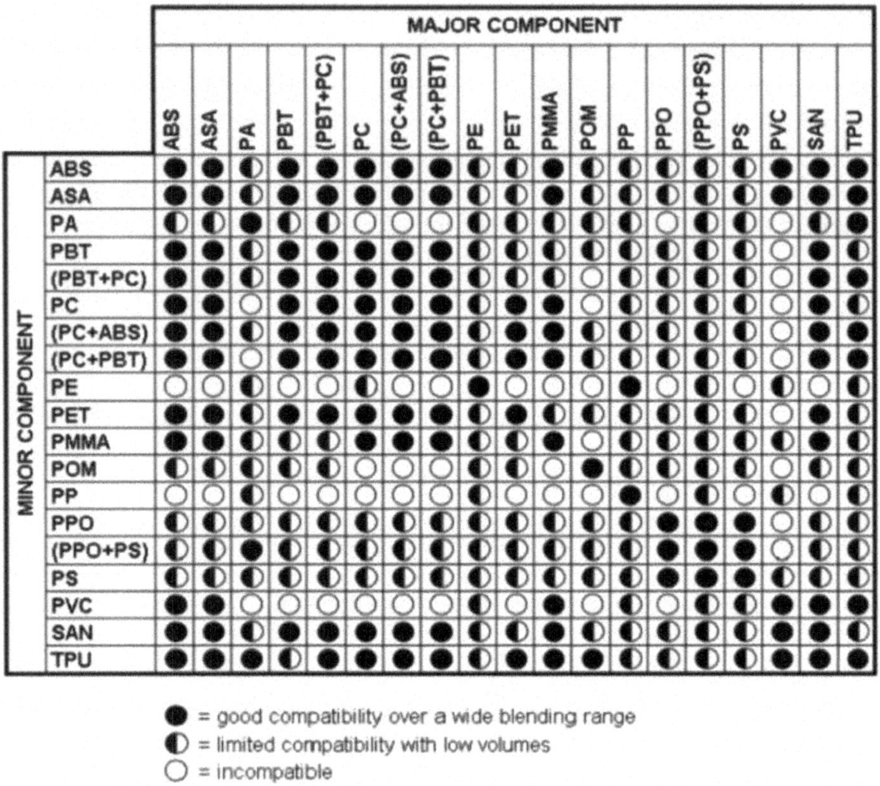

Fig. 4.1 Recycling compatibility of thermoplastics [3]

relevant when looking at Life Cycle Assessments that also conclude that from an environmental perspective dismantling is not clearly preferably when compared to post-shredder treatment separation that can separate more materials albeit at lower purity levels [6].

DfD is often not effective in practice because dismantling is influenced primarily by work organization/experience and more efficient waste processing alternatives exist.

Similarly, the DfR approach has been called into question by a series of studies. From an environmental perspective, for complex products, there is no evidence that mechanical recycling is environmentally favorable compared to other treatment options for plastic waste [6–8]. For example, when looking at automotive recycling and taking the vehicle's whole life cycle into account, the end-of-life phase of non-metals does not play any significant role in terms of potential environmental impacts or recycling credits, i.e. all these efforts result in no remarkable improvement for the environment [9–11]. The overall environmental performance can sometimes be jeopardized when optimizing the end-of-life vehicle treatment by changes in product development to the detriment of the overall performance (e.g. if lightweight composites are replaced by heavier, easier recyclable materials) [7, 10, 11]. Also, new technologies—including post-shredder-treatment separations based on near-infrared spectroscopy or x-ray separations—may limit the loss of quality for mixed recycling products, i.e. advancements in recycling technologies may be more effective than design restrictions in some cases.

In addition, it was increasingly clear that *limits to growth* are not predominantly determined by resource scarcity [2] but by the inability of the planet to cope with the emissions caused by human activity due to the ecological footprint of humanity overshooting all ecological limits [12, 13]. Resource scarcity has turned out to not be the primary concern for economic development either. In other words, "the stone age did not end because the world ran out of stones, and the oil age will not end because we run out of oil" [14]. This is mainly because of the development of backstop technologies, i.e. technologies substituting resources that are unavailable or excessively expensive.

Similar to the *limit to growth*, there is also a limit to circularity where increasing recycling rates beyond a certain point can lead to negative environmental impacts, i.e. when the marginal environmental negative impacts are higher than the reduced environmental impacts by substituting primary material [15].

DfR might be environmentally counter-productive where in conflict with designs that optimize the use phase (e.g. lightweighting), due to backstop approaches and recycling quality concerns.

Nevertheless, elements of Design for Disassembly and Design for Recycling can still be relevant. Firstly, usually the use of primary materials is often linked to higher environmental impacts than recycled materials (compare Fig. 8.4). Secondly, there are regulatory requirements in some cases. For example, the EU Commission is proposing in their revision of the packaging directive recycling-oriented design criteria. In addition, there are end-of-life sensitive products and materials where these guidelines are key to establishing a circular economy. Last but not least, these guidelines have positive impacts on repairability (in particular design for dismantling) which supports economic sustainability objectives as well. It must always be made clear that these guidelines have to be balanced and checked for the trade-offs in other sustainability areas.

Design for Disassembly and Design for Recycling are experiencing a revival—in particular, in the context of geopolitical considerations that require increasing robustness against supply chain disruptions to prevent one or few countries from monopolizing raw material supply. The EU Commission has established a list of so-called *critical raw materials* that is updated every three years to address this issue. The list is based on a review of economic importance considering the end-applications of these raw materials and supply chain disruption risk. For 2023, the EU Commission identified 34 critical raw materials including bauxite/aluminum, rare earth minerals, lithium, and cobalt alongside some strategic raw materials that do not meet these criteria (Cooper and Nickel) [16]. The EU Joint Research Center (JRC) recommends a mandatory dismantling for critical raw materials like rare earth minerals in e-motors and "from 1 January 2031, manufacturers shall ensure that the design of the vehicle and joining, fastening or sealing techniques do not prevent disassembly operations for repair and reuse purposes of e-drive motor, when present. Manufacturers shall also ensure that any software component controlling the operation of the e-drive motor does not impede its normal functioning after repair and reuse operations". This should be ensured by manufacturers by providing instructions on the disassembly operations that include:

- The type of operations to be performed.
- The type and number of fastening technique(s) to be unlocked.
- The tool(s) required [17].

In other words, the requirement consists of:

- Classical Design for Disassembly rules.
- Software compatibility requirements to enable future reuse.
- Information requirements for the manufacturers towards end-of-life operators.

Revival of DfD and DfR is likely due to the sustainability impacts of replaced virgin materials, regulatory pressure, virgin material supply disruption risks, and better repairability.

In the automotive industry, there is a precedent of ensuring an information flow to the end-of-life operators in the form of the International Dismantling Information System (IDIS) that provides dismantling recommendations such as the required tools to the dismantlers [18].

Finally, Design for Recycling or Design for Disassembly has been broadened to encompass all aspects of the circular economy based on the various strategies (Table 4.2 covering technical and biological approaches).

There are numerous approaches to foster a circular economy by design. "Circular economy" is defined by ISO/DIS 59004(en) as an economic system that "uses a systemic approach to maintain a circular flow of resources, by recovering, retaining or adding to their value while contributing to sustainable development" [22]. In the related ISO/DIS 59020 [23], different principles and examples for measuring indicators are listed. Currently, several indicators have been developed to assess the circularity of a product, including:

Table 4.2 Circular strategies—R10 Model [19] extended to nature-inspired design

Principle	R#	R-strategy	Description
Smarter product use and manufacture	R0	Refuse	Make product redundant by abandoning its function or by offering the same function with a radical different product or service
	R1	Rethink	Change product design, service design, and user behavior. Make product use more intensive, e.g. by sharing models or by multifunctionality
	R2	Reduce	Consume fewer technical or natural resources, increase efficiency in production or use
Extend lifespan of product and its parts—slow down material flows	R3	Reuse	Reuse of the functionality of product by another or same consumer for same or other purposes
	R4	Repair	Repair and maintenance of products to preserve the original function
	R5	Refurbish	Restore an old product and bring it up to date
	R6	Remanufacture	Use parts or all the discarded products in a (new/renewed) product of the same function
	R7	Repurpose	Use parts or all the discarded products in a new product of a different function
Useful application of materials	R8	Recycle	Process discarded materials to obtain a material of same or lower quality
	R9	Recover energy	Process discarded materials to obtain energy
Nature inspired design	R10/ N1	Regenerative	Biological materials are used in open loops with waste being biodegraded into nutrients. Technical materials stay in closed loops (cradle-to-cradle) [20]
	N2	Biomimicry	Design analogous to biology [21]—like packaging such as bananas or eatable plates

For modification © Wulf-Peter Schmidt 2024. All Rights Reserved

- The *Material Circularity Indicator* (MCI): assesses the share of materials that are recycled taking into account bio-based, renewable materials, as well as finite resources and materials. It can be applied to a product, a company, or an industry and considers the type and quantity of materials as well as their origin and destination. The MCI is based on the Linear Flow Index (LFI) that is defined as the share of the primary resources, so-called virgin materials (V), flowing linearly (without any recycling, reuse, remanufacturing, or other secondary resources), and the mass of unrecoverable materials (W) to the total material flow. The authors consider renewable materials as non-virgin in this LFI context. The LFI can have any value between and including 0 and 1. The MCI is basically the complementary indicator; however, the MCI additionally considers a utility factor F(X) that accounts for the lifetime and intensity of use of a product relative to the industry average.

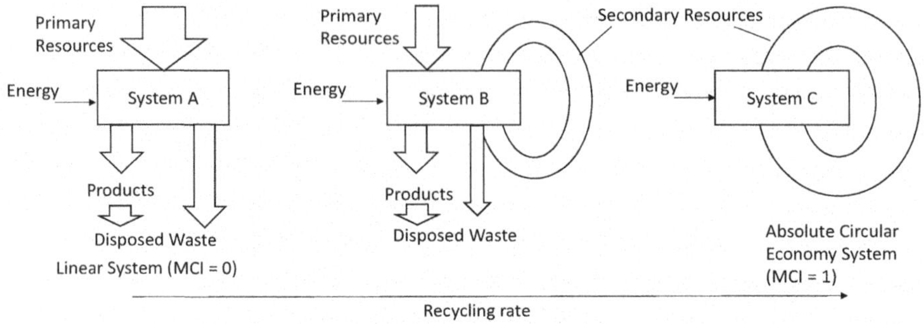

Fig. 4.2 Material circularity index and recycling efficiency taken to extremes—efficiency of overall system at high recycling rates not taken into account. © Wulf-Peter Schmidt 2024. All Rights Reserved

This is meant to acknowledge the positive impact of durability/utility design on the circularity of the product [24].

- Regulations also scrutinize *recycling efficiency* which is defined, for example, in the EU Battery Regulation, as the "ratio obtained by dividing the mass of output fractions accounting for recycling by the mass of the waste batteries input fraction, expressed as a percentage". This is an important metric of efficiency on a recycling process level. This definition does not differentiate between the different qualities of recycling output.

  The assumption is that an increase in recycling rate or in the Material Circularity Index is always preferable. This discounts the potential risks of an inversely proportional relation between recycling rate and efficiency (Fig. 4.2).
- *Social circularity* indicators are used, for example, in Social Life Cycle Assessment and various approaches can be used as shown inTable 2.11.
- The so-called *Material Design for Sustainability* (MDS) indicator focuses on the 3 R's reduce, recycle, and recover [25].
- From an environmental Life Cycle Assessment perspective, a *circular footprint* can be established (see section Circular Footprint Formula Approach).

> Various circularity metrics have been proposed. It is important to assess how meaningful they are taking into consideration efficiency, system boundaries, and sustainability impacts.

- *Resource inflow circularity* indicators represent the circularity of resources that flow through the system boundary into the studied system [23]. All resource inflows—besides water and energy—should be accounted for to calculate the content of the product composed of recycled material, reused portions, virgin renewable material, and virgin non-renewable materials respectively. These four types of content should add up to 100%. The calculation can also be made from average numbers. The system boundaries should be established similar to a Life Cycle Assessment (see Sect. 2.2.1).

- *Resource outflows circularity indicators* reflect the "circularity performance of resources that flow out of the boundary of the system. All resource outflows, including secondary materials produced .... except water and energy .... should be accounted for in this category" [23]—again looking at the system boundaries. Possible indicators reflect:
  - The percentage of actual reused content derived from an outflow X.
  - The actual recycled content is derived from an outflow Y.
  - The percentage of an outflow Z that actually goes into a "renewable recirculation" or to the biological cycle.
  - The remaining non-circular flow is seen as a loss caused by the considered system.

These percentages should add up to 100%. For the outflow Z, "the returns to the biosphere can be described as a fully biodegraded material or substance that microorganisms can decompose and that degrades to organic molecules that living systems can use further, for example via composting and anaerobic digestion" [23].

One challenge in deriving the actual shares is to acquire reliable information about what actually happens to the outflows outside the system boundaries. For example, a sorting process may separate recyclable plastic streams that are exported to a recycling operation on another continent where the information flow might be limited. Also, the quality of later reuse or recycling products is usually not reflected in these shares.

These metrices are also targeted in regulations. For example, the EU End-of-life Vehicle Regulation (formerly the EU End-of-life Vehicle Directive) mandates a reuse and recycling rate of 85% and a recovery rate (i.e. reuse+recycling+energy recovery) of 95% for each EU Member State. Interestingly, these rates can exceed 100%. The reason is the definition of recycling and recovery rates on a national level. The rates are calculated from the average weight of the end-of-life vehicles (ELVs) per year. From 2010 to 2014, the official, actual recovery rate exceeded 100% in Germany. The reason was a scrappage scheme that was introduced to subsidize new vehicle sales as a reaction to the 2009 financial crisis. Suddenly ELV operators had to accept and store more ELVs per year than they could process in that same year. For several years this meant that ELV operators processed more ELVs than they received in that specific year (Fig. 4.3).

Other circularity indicators include:

- *Energy circularity* indicators reflect the respective share of energy derived from renewable energy sources, residual, non-renewable resources, as well as virgin (primary) non-renewable energy sources [23].
- *Water circularity* indicators are still evolving and can include various aspects [23]. One example considers the "percentage of annual water demand that is derived from circular sources". These circular sources represent for example
  - recycled water

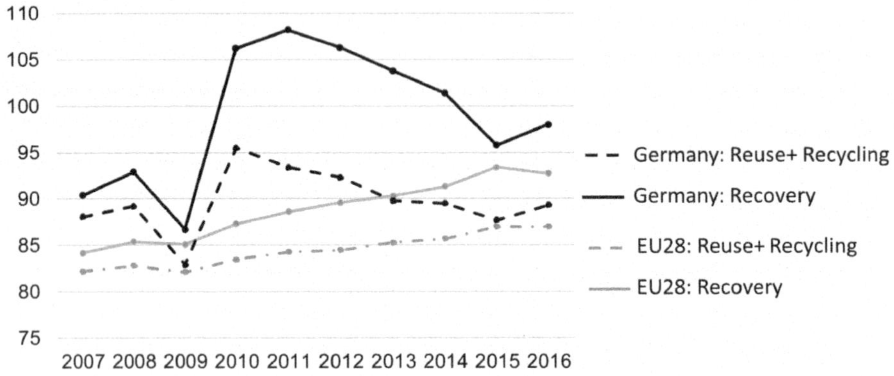

Fig. 4.3 Historical end-of-life vehicle recycling and recovery rates in Germany and EU28 (based on eurostat data). © Wulf-Peter Schmidt 2024. All Rights Reserved

  - non-freshwater natural sources such as sea or brackish water,
  - product/commodity embedded water (e.g. food and beverage industry),
  - "rapidly" renewable freshwater sources, such as surface water from areas that are not water-stressed,
  - renewable groundwater
  - harvested rainwater that is completely renewed by precipitation and
  - natural flows.

  In addition, these circular sources must be subject to strong governance with equitable and sustainable allocation to all users. Also, an approach where water is treated and then discharged back into the water source is preferred.
- Economic circularity indicators like the Revenue Share of Circular Resources (or products) (RSCR) measure the share of an organization's total sales revenue generated by the use of circular resources or products. It is expressed as the % of sales revenue generated per year through the use of circular resources [23].

  Overall, metrices for circularity face different challenges:

- The quality of circularity is either not taken into account or difficult to measure,
- High circularity values are not necessarily better than low values (due to efficiency, artifact of temporal or geographical system boundaries, sustainability impacts, etc.),
- Balance between simplicity to keep the metric easy to apply and complexity to incorporate different qualifiers to make the metric meaningful.

4.1.2 *Design for Environment and Eco-Design*

Knowledge about potential trade-offs on a more holistic product life cycle level has led to the intention to integrate an environmental life cycle thinking approach into product development [26].

In consequence, there are numerous Design for Environment (DfE) tools (e.g. Environment-Failure Mode Effect Analysis, Multi-Criteria-Requirement Matrix, and checklists, simplified life cycle tools, or comprehensive material selection tools [27]). For all these tools, the holistic approach is important—they cover not only one but all-important environmental impacts relevant to the product.

> DfE tools cover several environmental impacts along the entire life cycle. Examples include E-Failure Mode Effect Analysis, Multi-Criteria-Requirement Matrices, Simplified LCAs.

The international standard ISO/TR 14062 on "Integrating environmental aspects into product design and development" has been withdrawn. Various tools and methods have been proposed and tested, but very few tools and methods are used in day-to-day practice. It is important to identify those tools that apply most to company-specific general product development tools, management systems, approaches, and company culture and fully integrate these tools into everyday practice [26].

4.1.2.1 Integrating Environmental Aspects in Traditional Design Tools

An example of this is the Environment Failure Mode Effect Analysis (E-FMEA). This method can be extended to encompass all sustainability criteria where appropriate. A classical Potential Failure Mode and Effects Analysis in design (D-FMEA) or in manufacturing and assembly processes (P-FMEA) is 'a systemized group of activities intended to: (a) recognize and evaluate the potential failure of a product/process and its effects, (b) identify actions which could eliminate or reduce the chance of the potential failure occurring, and (c) document the process. It is complementary to the design process of defining positively what a design must do to satisfy the customer' [28]. This quality tool is widely used in the aviation and automotive industries and several variations exist for other areas like Hardware FMEAs, Software FMEAs, System FMEA, etc. It usually evaluates the severity and likelihood or occurrence of failures as well as how easily the failures can be detected.

> Integrating sustainability in product development is best done by using existing design tools already used in the organization and complementing these tools. An example is FMEAs.

Table 4.3 Environment FMEA based on activities along the life cycle [1]

ENVIRONMENTAL FMEA DESIGN

Customer	Part name.	Dwg No.	Supplier
	Spark		
Function	**Date**	**Issued by**	**Project** / **Issue**
Washing machine	1996-08-09		PW111 /

No	Lifecycle Phase	Activity	Environmental aspect	Environmental impact	S	I	A	EPN/F	Recommendations	Decisions taken	S	I	A	EPN/F	Dept/Sign
	ENVIRONMENTAL CHARACTERISTICS				RATING				ACTION-STATUS						
	Production	Welding of 25 different pieces	Emissions to air from welding the drum	Toxical	2	1	3	6/5	Single-shot polypropylene moulding the drum						
		Cleaning and painting the cabinet	Emissions to atmosphere from VOC	Toxical and ozon depletion	2	1	2	5/3	Look into the possibility for prepainted cabinet						
		Working the components for the drum	Metal waste from working 30 different components	Resource consumption material	1	1	1	3/7							
	Use	Washing and rinsing	Use of resources water (150l/wash)	Resource consumption water	1	2	1	4/5							
		Use of detergents	Release to water (waste water)	Eutrophication	1	2	1	4/3							
		Pumpning and heating the water, spinning the drum	Use of resources electricity (1,50 kWh)	Resource consumption energy 50% nuclear power	2	2	2	6/4	Develop electronic control system to regulate energy use						
	Disposal	Disassembly of polymer components	Use of materials polymer (not marked)	Resource depletion	2	1	1	4/4							
			Emission to atmosphere from flameretardents	Ozone depletion	3	2	2	7/5	Does the plastic material contain brominated subst.?						
		Disassembly of electronic components	Contamination of land from toxic materials e.g. cadmium etc	Toxical	3	2	2	7/8	Does the plastic components contain toxic materials?						

An adjustment of the general FMEA approach for environment or sustainability criteria should be straightforward if focused on recognizing and evaluating potential negative environmental or social impacts. Different proposals for an Environmental FMEA have been made. A Swedish approach starts with activities along the life cycle, then determines and evaluates their environmental impact and finally concludes with recommendations [1, 29]—see Table 4.3.

An approach closer to the classical FMEA looks at the environmental objectives and analyzes how a failure to achieve these objectives might happen, what the likelihood or occurrence of these failures might be, and how easily these failures can be detected. It concludes with mitigating recommendations [30]—see Table 4.4.

These tools are traditionally focused on environmental aspects, however, they can and should be extended to also cover social and economic aspects and objectives.

4.1.2.2 Design for Durability in Eco-Design

Durability or Design for Durability is the most important eco-design requirement named by companies in a survey in Germany published in 2022. This was followed by lower resource consumption and lower environmental impact [31]. A potential

Table 4.4 Environment FMEA close to classical FMEAs [30]

Environment - Potential Failure Mode and Effect Analysis (Environment - FMEA)

___ System
_x_ Subsystem: MAC
___ Component
Model Year / Vehicle(s): T139
Core Team: Climate Control

Design Responsibility: Muy Li
Key Date:

E-FMEA Number:
Page 1 of 2
Prepared by:
FMEA Date (Orig.): (Rev.):

Objectives (Examples)	**Potential Failure Mode** (Examples)	**Potential Effects of Failure**	**Sev**	**Class**	**Potential Causes / Mechanisms of Failure**	**Occur**	**Current Test Methods**	**Detec**	**R. P. N.**	**Re-commended Action(s)**	**Responsibility & Target Completion Date**	**Actions Taken**
Use of renewable or recycled material, no restricted substance or materials listed with high environmental footprint	*None used / less than benchmark; substance x used*	*environmental footprint cradle-to-gate worse than target*	*7*	*EC*	*Material properties of alternatives are not in specification*	*8*	*International Material Data System.*	*8*	*448*	*Materials engineer to check spec., benchmark options*	*Engineer Smith* *Next E-FMEA meeting*	
Energy consumption below x (low friction, lowest aerodynamic, low consumption of auxiliaries, …)	*Energy consumption above x due to auxiliaries*	*Increased gap between tested and real world energy consumption*	*10*	*ES*	*Carry-over auxiliary with legacy efficiency*	*10*	*Test with and without auxiliaries*	*8*	*800*	*Check feasibility of new auxiliary*	*Finance manager Bull* *Next E-FMEA meeting*	
Non-destructive dismantling possible within x seconds, no material complexity	*Too many different materials, too many different fasteners, etc.*	*Increased insurance costs* *Longer destruction time if not experienced*	*3*	*ES*	*Different types of fasteners used*	*5*	*Stopping time*	*8*	*120*	*Component engineer to check feasibility of harmonizing fasteners*	*Engineer Li* *Next E-FMEA meeting*	

SEV: Severity. Assessed from 1 (low) to 10 (Immediate serious Environmental Risk). Rating definition by company
EC: Environmental Compliance Class, ES: Environmentally Serious
Occurrence: 1 (low) to 10 (high)
Detection: (1) Test suitable, available, applied by correct people at best time to (10) test methods will detect the potential cause or resultant failure.

metric for durability is the "average lifetime of a product or material relative to the industry average". This metric is established by considering the reliability of the product or material and the lifetime extensions from maintenance, repair, updates, upgrades, and where applicable refurbishment throughout its expected lifetime [23].

The idea of Design for Durability is to slow down resource flows. This is a holistic eco-design strategy especially applicable to production-sensitive products and services where a longer product life "capitalizes" on the "investment" of the cradle-to-gate environmental impacts for a longer period of time. From an environmental life cycle perspective, the optimal life span can be calculated by comparing it with alternatives like substitution with a new but low-impact product [32].

> Design for Durability is the most used eco-design approach. However, for non-production-sensitive products, the environmental optimal product lifetime has to be checked.

In the survey, there is a remarkable variation in the importance of *repairability*. It is seen as the second most important eco-design criterion by producers that are also service providers while producer-only companies deprioritize this criterion—potentially to maximize their sales [31]. This sheds some light on the motivation of different types of companies. Repairability can, of course, also extend the product's lifetime and durability.

4.1.2.3 Checklists and Multi-criteria-requirement Matrix

A simplified life cycle thinking approach can be done somewhat qualitatively in the form of a checklist or matrix approach. Along the life cycle, different environmental, economic and/or social aspects can be systematically considered. Usually, the same aspects are checked for each life cycle phase. Several approaches including the Matrix Element Checklist for ERP, MET-Matrix (Materials, Energy, Toxic emissions), Environmentally Responsible Product/Process Assessment Matrix, and Environmental Product Life Cycle Matrix have been used [33]. The aspects covered and the weighting of these aspects is selective and subjective. On the other hand, it is a simple approach that includes sustainability issues considered relevant in decision support. It is important that the chosen weighting and selected criteria are proportionate to their significance along the life cycle. The shortfalls and advantages are obvious when looking at the approach shown in Table 4.5.

> Environmental Multi-Criteria-Requirements Matrix is a semi-quantitative, simplified life cycle tool based on very few environmental aspects along the life cycle and subjective weightings.

Table 4.5 Example for an Environmental multi-criteria-requirement matrix

Raw materials		Option A	Option B	Option C	Score range
Product	Contains listed substance of concern[a] or other stringently regulated substance				1–3 no issue 4–6 coming issue 7–10 restricted-ban
Process	Substance of concern[a] or other stringently regulated substance are used in manufacturing/process operations prior to own control (e.g. SF6 for magnesium production, CFC cleaning)				1–3 no issue 4–6 coming issue/image problem 7–10 restricted (7)—legally banned (10)
Waste	Type of waste				[b]
Energy	Energy used to acquire & manufacture raw material inputs (if data is available for each)				3–7 (3 lowest figure of the alternatives)
Water	Water used per unit production				1–4 (1 reduction from current, 2 status)
Raw Materials Total					
Raw Materials Total Weight Factor		2	2	2	
Raw Materials Total, weighted					Range: 10–82
Manufacturing and Assembly		Option A	Option B	Option C	Score range
Product	Applies/adds listed substance of concern[a] or other stringently regulated substance				See above
Process	Substance of concern[a] or other stringently regulated substances are used or generated				See above
Waste	Amount and Type of waste				[b]
Energy	Energy used per unit production				See above
Water	Water used per unit production				See above
Manufacturing and Assembly Total					
Manufacturing and Assembly Total Weight Factor		2	2	2	
Manufacturing and Assembly Total, weighted					Range: 10–82

(continued)

Table 4.5 (continued)

Raw materials		Option A	Option B	Option C	Score range
System Use		Option A	Option B	Option C	Score range
Energy	Energy required to use a product over lifetime (electricity consumption over 10 years of use or driving a vehicle over 150,000 miles)				1–10 relative energy demand (current: 5)
Maintenance/ Operation	Use of substances of concern[a] or other stringently regulated substances or if applicable evaporation (e.g. interior material: smell, VOC; R134 emission, etc.)				1–3 no issue 4–6 coming issue 7–10 restricted (7)—legally banned (10)
System Use Total					
System Use Total Weight Factor		10	10	10	
System Use Total, weighted					Range: 20–200
End-of-Life		Option A	Option B	Option C	Score range
Waste	Type of waste (recyclable, landfill, incineration)				[b]
	Effort for additional treatment (energy, etc.)				3–7 (3 lowest figure of the alternatives)
Dismantling	Easiness of dismantling (if necessary)				[c]
End-of-Life Total					
End-of-Life Total Weight Factor		2	2	2	
End-of-Life Total, weighted					Range: 6 (4)–54
Design Total					

[a]List of substances of concern as per company or regulation
[b]1–2 returnable or easy recyclable waste, 3–4 energy recovery, 5–6 normal landfilling, 8–10 hazardous waste (higher score for a bigger amount)
[c]1–2: w/o tools, 3–4 with tools, 5–7: special tools needed (higher score for time needed), 8–10: not possible
© Wulf-Peter Schmidt 2024. All Rights Reserved

4.1.3 *Design for Sustainability*

However, it turned out that the holistic environmental perspective was also unsatisfactory. Products that are designed for a low environmental impact—or even better for a positive environmental impact—are not automatically also sustainable. The "greenest" products will fail if, for example, they are not affordable or if the social implications are ignored. Aspects beyond environmental issues needed to be included in the tools systematically.

The first logical step was the extension of the environmental life cycle tools to cover economic aspects as well. Life Cycle Costing is an ideal tool for this (Sect. 2. 3). However, a more comprehensive approach was needed to reflect the key product attributes that contribute to all the dimensions of sustainability.

Another approach is to include social product design principles in whatever way is possible. Tools like Design for the Base of the Pyramid (BoP) exist that focus (in contrast to the Top of the Pyramid approach) on low-income consumers or countries as consumers or producers and business partners. This can be achieved by product co-creation with such communities accompanied by sourcing in these areas, etc. [34]. However, these approaches are rather isolated, suffer from various issues (lack of information about needs and capabilities, inadequate infrastructure or access to finance and talents to transfer consumers to micro-producers giving them income streams to be able to consume), and lack a holistic life cycle perspective.

Design for Sustainability approaches is another logical evolution as they include all relevant environmental, social, and economic aspects. However, they need to balance the desire to cover everything comprehensively with the constraints of their practical implementation. Ideally, those who design or construct the product/part apply Design for Sustainability thinking to their product without the need for additional training/expertise in sustainability aspects and without slowing down the design process. This increases the likelihood these techniques being used in real-life and also improves understanding, ownership and acceptability of sustainability targets. An example of this is to include social and economic aspects in Multi-Criteria-Requirement Matrices or FMEAs.

> Design for Sustainability (DfS) covers environmental, social, and economic aspects along the life cycle. DfS should not be too complicated to ensure applicability in practice.

4.2 System-Oriented Design for X

Another evolution is the movement away from designing with a focus on a product and towards a wider system design approach. The idea is to overcome the limitations of the focus on the product such as not sufficiently leveraging the engagement of other stakeholders. For example, producing remanufactured smartphones can only be successful if:

- Old products like smartphones are returned to the manufacturer or someone who can remanufacture the product. There are an estimated 700 million unused smartphones in European homes with an estimated recycling value of 1 billion Euros [35]. It is unlikely that many of these phones are still being used in their original or in new innovative ways [36].
- Customers accept remanufactured products. This requires trust in the quality of the remanufactured product, compatibility with the latest technology, guarantees, etc.

Therefore, product design must look at four areas—business model innovation, product design, service design, and supply chain management [37]:

- A circular business model closes, slows down, intensifies, dematerializes, or narrows resource loops. Durability and repairability of products are still considered as they were in classical, product-focused design. However, the focus is now on how to ensure that these products are used longer or reused in real-life by creating sharing platforms, product-as-a-service models, new ownership approaches, or other approaches. Another approach is the Emotionally durable design (EDD) that aims to enhance the durability of the relationship established between users and products [38].
- Product design that supports other design for x approaches as shown in Table 4.1 such as Design for upgradability and adaptability, reliability, modular design, etc.
- Service designs should ensure that after-sales services allow product upgrades and effective repair for long-term use to ensure customers keep their current product instead of looking for a new product. The design should support "performance models" where the service provider is responsible for ensuring the product continues to fulfill a certain function as required by a service contract or "access models" where product ownership is not with the customer who pays for access to the product service.
- Supply and value chain management should actively enable circularity by ensuring adequate information flows to relevant life cycle stakeholders, ease of transitioning from one to the next step in the circular processes, ensuring material flows for recycled materials, etc.

System-oriented design reinvents business model, product/service design, and supply chain management to keep products in use for longer. It also closes material/information loops (Fig. 4.4).

The above-mentioned emotionally durable design (EDD) tries to keep users bonded to the old, current product by always creating novelty and keeping functionality up-to-date with software updates (ideally free, without hassle). Simultaneously, the product continues to evoke good memories and pride in the product (e.g. by personalization). Obviously, this physically requires a very durable and repairable design but also timeless esthetics and durable, functional adaptability. This is difficult to design from scratch but is seen in any iconic product (certain watches, kitchen devices, cars, even fashion, and many more products).

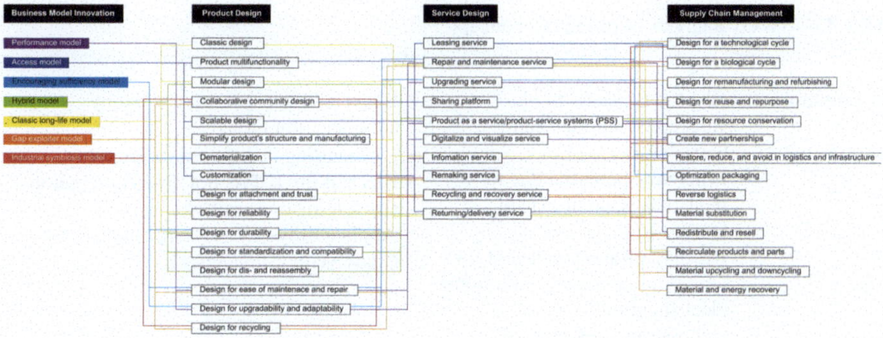

Fig. 4.4 Seven circularity design characteristics of business models for the electrical and electronics industry [37]

An example is the Re-factory concept [39] where a division called "Re-Trofit" offers the upgrading of conventional, used vehicles to electric vehicles as well as the refurbishing of second-hand vehicles. The division "Re-cycle" targets the dismantling of end-of-life vehicles at an industrial scale by reversing the assembly line for dismantling. The above-mentioned issues (getting old products to manufacturer and customer acceptance for remanufactured products) are partly addressed by working together with dealers who send the old vehicles (not end-of-life vehicles) to the producer who provides guarantees and a faster turnaround due to the high availability of spare parts (from dismantling and primary spare parts). The automotive manufacturer (OEM) uses a manufacturing site for the Re-factory division that otherwise would have to be shut down. There are strengths and weaknesses in the Re-factory concept (see Table 4.6). It can be successful if the OEM can ensure a very lean and efficient implementation and is able to squeeze the existing end-of-life operators out of the market.

4.3 Product-Service Systems

Similar to system-oriented designs (Sect. 1.2), a Product-Service System (PSS) is a type of product development that goes beyond the product itself. However, here the focus is on selling a service instead of or in addition to selling the product ownership. Various approaches exist, including [40]:

- Services providing added-value to the product life cycle. In addition to the product, an additional service is also offered to the customer. This service can, for example, guarantee functionality and durability (i.e. product life extension) of the product. Service contracts could include maintenance, repair, upgrading, and substitution services over a specified period of time (for example, more power or connectivity for a vehicle)—and potentially a returning of the product after this period of time.

Table 4.6 SWOT analysis of re-factory concept

	Strengths	Weaknesses
Opportunities	• OEM gets access to old vehicles, materials, and batteries • Efficiency potential due to industrial scale of existing facility • Keeping products relevant for longer by upgrading emissions/ electrification • Generating knowledge for product development	• Recycled materials are needed by suppliers, not OEM • Dismantling is different to assembly • Electrifying ICE is not always an option (packaging constraints, costs) • Applicability of knowledge unclear for future vehicle concepts
Threats	• Logistical challenges can be addressed by size/presence of OEM in region • New business areas can generate synergies and revenues • Customer acceptance of remanufactured/dismantled parts increases with brand strength	• Many re-factories needed to limit maximal distance to old cars collection points/ dealers • Management distraction and diversion of investments beyond core business • Agile end-of-life operators compete with the OEM Re-factory concept at lower fixed costs and at shorter distances

© Wulf-Peter Schmidt 2024. All Rights Reserved

This approach requires a design that minimizes costs through better reliability as well as the ability of maintenance, repair, and upgrading (and disposal).

- Services providing final results to customers. Here, the product ownership remains with the producer/service provider who offers a customized mix of services (as a substitute for the purchase and use of products), in order to provide a specific final result (in other words, an integrated solution to meet the customer's needs). The service provider is paid for providing the agreed upon results (e.g. thermal comfort). The customer is freed from any problems linked to ownership (e.g. in agriculture a pest control service instead of selling pesticides or in industry a sale of thermal comfort instead of heating systems or m^2 of painted surface instead of selling paint ➔ out-sourcing services). This approach requires a design that minimizes the energy and materials consumed in use. In some cases, reliability, maintenance, repair, and upgrading might also be important. Economies of scale can often be established easier.
- Service-enabling platforms for customers. A platform is offered to customers to access products and services. The customers do not need to own these products and are only paying for the time the product is in use (leasing, pooling, sharing certain goods). This enables access to scarce products (e.g. co-operatives purchasing tractors in agriculture or a pool of computers and other office equipment or offices OR allowing access to certain products in developing countries). This requires a design that is robust/easy to repair (multi-users), flexible (for different purposes), and durable.

Product-Service Systems are based on services instead of/in addition to product ownership or utilizing sharing platforms. Supporting design focuses on durability, repair, and consumption...

4.4 Design for Sustainable Behavior

The Design for Sustainable Behavior (DfSB) method aims to create product designs that encourage a more sustainable behavior of users and with that improve the sustainability performance of products, services, and systems in their use phase. There are 45 or more DfSB methods [41] that are often based on:

- Customer/User research to understand the behaviors, attitudes, and motivations of users when using a product, identify potential needs of customers, and opportunities to change behaviors towards more sustainable practices.
- Identification of suitable behavioral change strategies from user research including:
  - Match—adapt products and services to current user behaviors but design in such a way that negative sustainability impacts are reduced.
  - Enlighten—motivate consumers by increasing their knowledge about sustainability implications of their actions by feedback mechanisms like showing real-time energy consumption in smart meters relative to benchmarks or educational approaches like labels, intensity of signals, and alerts.
  - Spur—encourage customers directly by gamification which creates a sense of competition, incentivizes better behavior, and utilizes social influence/social norms.
  - Steer—make sustainable behavior an obvious and easier choice through design. Examples include having default positions in an eco-mode, requiring additional effort to change or add an emotional appeal to the right choice while making it harder for users to follow unsustainable behaviors.
  - Force—make non-sustainable behavior impossible by offering only cleaner technologies, products, or services.
- Life cycle thinking which considers the sustainability impact of the product at each stage of its life cycle. This helps designers to identify opportunities to encourage more sustainable behaviors.
- Use of persuasive strategies in design. Examples include visual cues, social norms, and gamification to encourage sustainable behavior.

Design for Sustainable Behavior approaches work best if they support user satisfaction, achieve better user understanding, have obvious co-benefits for users, are perceived as innovative, can build on a solid trust basis, and are rightly timed, so happen where and when consumers are open for transitions. The biggest barriers include the diversity of consumer groups, the difficulty of changing behavior patterns and ensuring the changed behavior is maintained, time/resource/knowledge constraints and rebound effects.

Design for Sustainable Behavior aspires to a design that encourages consumers to adopt the desired sustainable behavior by feedback mechanisms, gamification, or design hints.

4.5 Design for Sustainable Systems

Design for social innovation for sustainability (DSIS) goes beyond the product and service level and aspires to assist with the conception, development, and scaling-up of sustainability innovation within communities and social networks. It consists of the following principles:

- Life Cycle Stakeholder analysis to identify their interests, needs, motivations, priorities, and perspectives. Based on this, the Life Cycle Stakeholder engagement can happen in such a way that is meaningful and relevant for all.
- Life Cycle thinking—identifies issues and solutions regarding sustainability along the life cycle.
- Systemic thinking—considers via root cause analysis the entire system in which a challenge exists, including its social, economic, and environmental dimensions, and looks for solutions that mimic natural ecosystems.
- Co-creation—development of designs and business models together with life cycle stakeholders, including communities and organizations.
- Participatory design—this can include workshops, focus groups, and other collaborative methods that help to ensure that solutions are user-centered and meet the needs of those who will use them.

Neighborship schemes where citizens help each other are practical examples of DSIS.

While the DSIS approach is often limited to local initiatives, a broader approach is the Design for System Innovations and Transitions that aims at bigger-scale transformation of socio-technical systems through (strategic) designs based on technological, social, organizational, and institutional innovations. This approach combines design for product-service systems that aim to transform production–consumption systems through business model innovation (see Sect. 1.3) and DSIS which aims to assist with social change. This combination allows for design decisions that consider the interrelationships between technologies, ecosystems, social and cultural practice, and city/national/regional governance [42]. Examples are the design of low-carbon and climate-resilient cities or regions.

Design for Sustainable Systems goes beyond products/services by creating whole ecosystems covering cities/regions that foster socio-technological innovations/system synergies.

The above-mentioned different design strategies have evolved from only considering individual products via product-service systems to complex systems covering regions (ideally leading to one global sustainable ecosystem). At the same time, the dimensions covered widened from individual design targets like repairability to cover one and finally all sustainability dimensions. With this evolution, the potential positive impacts of sustainable design increase significantly. The challenge is to keep these approaches practical and ensure that sufficient knowledge and talent are applied.

Design strategies evolved from a focus on products via product service to whole regions, from covering individual targets to covering one and finally all sustainability dimensions.

4.6 Examples for Design Strategies

4.6.1 Lightweighting

From a sustainability perspective, the general rule is "less is more" if it comes to the use of materials. For example, the smaller/lighter a vehicle is, the lower its negative impacts are as long as a similar type of material is used (see also Fig. 7.2). Alternative designs that use less material can be based on structural changes such as those used in bionic lightweighting. These include honeycomb structures, which provide more stability with less material [43].

However, a more holistic life cycle check is needed if lightweighting is based on replacing conventional materials with lighter materials that are linked to significantly higher impacts including higher energy demand. Therefore, automotive lightweighting based on lightweight materials cannot reduce the overall vehicle life cycle CO_2 and tailpipe CO_2 beyond 4–8 g/km. This is seen in Fig. 4.5 where theoretical changes caused by different lightweight materials are shown (high-strength steel (HSS), Aluminum (Al), Magnesium (Mg), and carbon-reinforced plastics). The

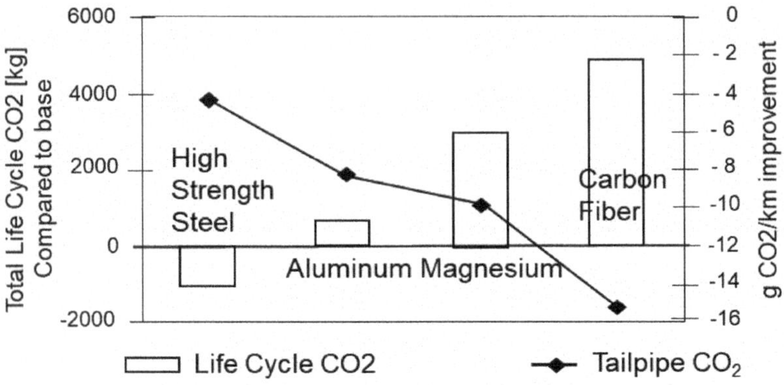

Fig. 4.5 Life Cycle CO_2 versus tailpipe CO_2 changes for different lightweight options. © Wulf-Peter Schmidt 2024. All Rights Reserved

figure displays the impact of material choice on homologated tailpipe CO_2 emissions (right y-axis, dark line) and total life cycle CO_2 performance (left y-axis, white bars).

Carbon fiber reinforced polymer composites (CFRP) have the potential to reduce weight by up to 65% when compared to steel automotive parts and up to 20% when replacing aluminum in aviation [44]. Their suitability depends on factors like cost, volume, part design, production times, experience, and environmental considerations. Figure 4.5 shows that the use of Carbon fiber reinforced polymers has the biggest potential to reduce tailpipe CO_2 emissions; however, the high-energy consumption and related emissions during production of CFRPs significantly increase the overall life cycle CO_2 emissions when compared to the steel reference case. Even if recycling of CFRPs is included, this option does not provide environmental benefits unless the production burdens (PB) of CFRPs can be improved or a significantly longer use phase than is common for Europe is assumed (Fig. 4.6).

> Using less of the same material leads to less pollution and social impacts. However, replacing conventional with lightweight materials requires a life cycle check for trade-offs.

4.6.2 *Selecting Low-Impact, Recycled, and Renewable Materials*

Incorporating sustainability considerations into material selection is a common approach to integrating sustainability into design. Often, the use of recycled or renewable materials is seen as a positive strategy, by definition.

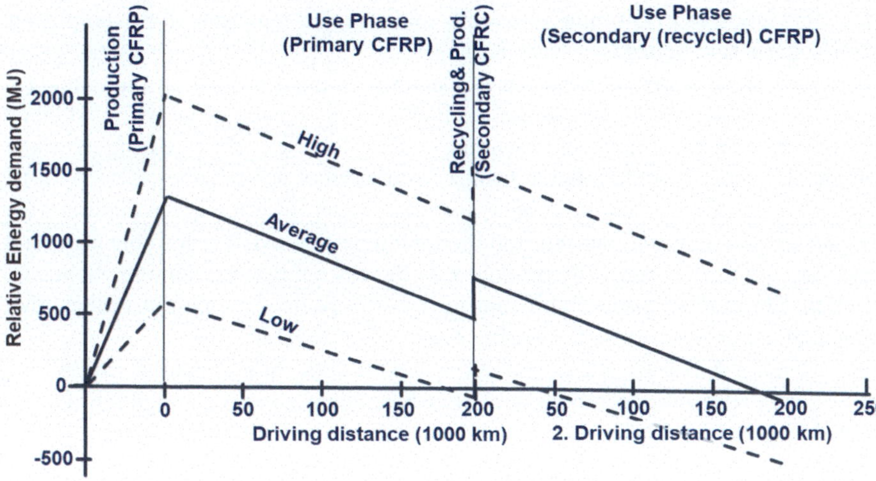

Fig. 4.6 Break-even envelope plot showing sensitivity analysis of life cycle energy demand. The envelope between high and low represents the potential impact values based on the range of values available in databases and the literature (based on [44] but corrected). *Note* The options for the recycling of fiber-reinforced plastics are detailed in Sect. 8.3

Renewable materials (such as hemp, flax, kenaf, jute, wood, and others) are defined as materials derived from plant, animal, or bacterial resources that are sustainably cultivated and regenerated within a human lifetime [45]. These materials are mainly used through fiber processing technologies and are often embedded in polymer matrices.

Recycled materials are based on the collection and processing of waste materials from two main sources [46]:

- Post-consumer recyclates (PCR) are based on waste materials that have passed through the hands of consumers. These waste materials are generated by households or by commercial, industrial, and institutional entities in their role as end-users of the product that can no longer be used for its intended purpose. It also includes the return of material from the distribution chain.
- Post-industrial recyclates (PIR) or pre-consumer recyclates are materials that are diverted from the waste stream during a manufacturing process. This excludes the reutilization of materials such as rework, regrind, or scrap that can be reclaimed within the same process that generated them. These waste streams should have been subjected to at least one additional process step.

Home scrap recycling (direct regranulation and re-feeding in the same facility) is excluded from recycled materials to avoid incentivizing process inefficiencies (see also Fig. 4.2). When counting recycled materials, only the weight of the recyclate itself (not the component as a whole) and only the recyclates actually used (not those that are only released or approved, but are not assembled) should be considered.

It is also possible to combine renewable and recycled materials, for example, by recycling harvest residues (see Sect. 8.3).

However, there is no one material that is always inherently better than another, as each material has negative impacts such as energy consumption or forced labor in harvesting of natural materials or impacts of collecting and processing of waste materials, i.e. there is no free lunch from a sustainability perspective.

For example, leather, natural rubber, and other materials often are in conflict with the do no significant harm criterion and should not be counted as renewable materials unless they are truly sustainably produced. In the case of leather, tanning chemicals and animal husbandry have significant negative impacts. For natural rubber, risk factors include unsustainable harvesting rates and child labor.

For recycled materials, the negative impacts of collection, cleaning, and reprocessing can be a limiting factor, particularly when it comes to producing high-quality recycled material.

> No material is always preferred, renewable/recycled materials may have worse impacts than conventional materials. Life cycle studies need to identify the best option for each application.

Therefore, a life cycle approach in material selection is required to identify materials that have lower life cycle impacts than alternative materials. This comparative life cycle approach considers all the processes required to achieve the required material properties compared to the substituted material. The higher the negative impact of the substituted conventional material (such as metals) the greater the likelihood that renewable or recycled materials are the better option, as long as they provide the same or better properties. There are several holistic material selection methods [47].

4.6.3 Substituting Restricted Substances or Materials

Substituting restricted substances is usually driven by regulations (e.g. EU Chemical regulation REACH) or customers (e.g. demand for allergy-free products). An example of this is the substitution of ozone-depleting (e.g. chlorofluorocarbons—CFCs historically used in air conditioning systems or sprays) or powerful greenhouse gases (e.g. hydrofluorocarbons—HFCs used in air conditioning systems or sulfur hexafluoride, the strongest greenhouse gas, used as a protection gas in the production of magnesium).

When substituting substances, a thorough analysis of trade-offs and other negative side impacts/do-not-significantly-harm criteria is required. Otherwise, the substitution efforts must be repeated which creates costly system changes each time affecting not only the substance but potentially also the whole product-service system. An example where this failed is refrigerants. CFCs were replaced by HFCs high greenhouse gas potentials (like R134a), then by HFCs with lower greenhouse gas potential (like R1234yf), then blended refrigerants, and finally CO_2. Each time, product designs and/or aftermarket systems and infrastructures needed to be replaced to support the substitution of the old refrigerant system by the next one. Where technically possible and affordable, a leapfrogging approach is best.

Similarly, substituting high-impact materials with low-impact materials is an approach that needs to consider the whole life cycle, i.e. should ensure that the new material is beneficial from a sustainability perspective along the whole life cycle. Reducing the negative impact in one life cycle phase or one aspect while creating new negative impacts in another life cycle phase or another sustainability aspect does not provide a long-term solution.

Rare earth minerals are an example of critical raw materials that are linked to environmental issues (ecotoxicity in mining, energy intensity in metallization, etc.) and have monopolized supply structures. Nevertheless, they are key for transformation as they are used for magnets in e-motors for Electric Vehicles or wind power generation. Alternative materials for rare earth minerals such as Barium Ferrit exist. These have neither the supply risk nor the cost issues of rare earth minerals. However, their properties have to be adjusted to the different applications [48].

> Substituting substances or materials needs to include a thorough Life Cycle check for all alternatives, looking at all dimensions, leapfrogging towards solutions with least trade-offs.

4.6.4 Applying Life Cycle Tools in Design—Product Sustainability Index

A good place to start a design for sustainability study is to identify potentially relevant issues and opportunities along the product life cycle from environmental, social, and economic perspectives. It is important that non-Life Cycle experts include life cycle thinking in their daily work as well. Simplified Life Cycle approaches allow product engineers to understand the implications of their design alternatives directly.

For example, the "Product Sustainability Index" (PSI) [49, 50] is limited to eight key performance indicators, including—for the case of automotive products—the three life cycle indicators *Life Cycle Global Warming, Life Cycle Air Quality* (Photochemical Ozone Depletion Potential) and *Life Cycle Cost of Ownership.* The eight key performance indicators selected should reflect those most relevant and material for the product (development) in question (here: automotive). A PSI for other products may have other key performance indicators. No weighting is applied between these eight indicators, instead they are shown independently in a spider diagram with a rating that is always based on the same principle:

- Zero percent amounts to the worst, current product of the manufacturer applying the Product Sustainability Index.
- Eighty percent is the best product currently on the market (all segments).

This normalizes the different units of the eight Product Sustainability Index indicators to the same scale. In addition, the scaling definition stresses that no current product is sustainable enough and further improvements are required. By constantly updating the zero and eighty percent benchmarks, a continuous improvement approach is integrated into the concept.

The Product Sustainability Index can be applied to all products based on the principles of materiality, no weighting, normalization based on benchmarks, and continuous improvement (Fig. 4.7).

The calculation of the Product Sustainability Index is done by a simplified Excel tool into which Vehicle Integration Engineers (for the vehicle level) or the Component Engineers (if applied on component level) enter only a few parameters that are known by these engineers like weight of the vehicle or component, fuel consumption and emission level, air conditioning, and any significant lightweight action. The Excel tool then uses a simplified Life Cycle vehicle model based on prior vehicle models to establish the Product Sustainability Index result. When compared to a full Life Cycle Assessment based on expert tools and an independent peer review, this approach showed very similar results with the difference being below two percent [50].

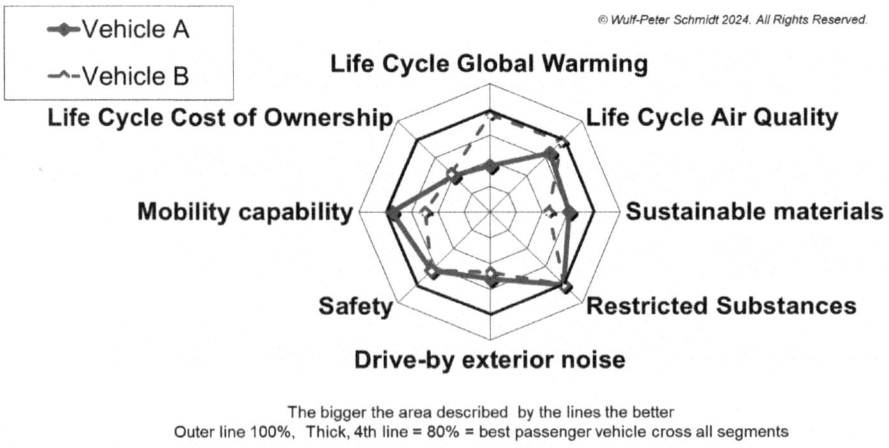

Fig. 4.7 Product sustainability index

4.6.5 *Environmental Business Case with Life Cycle Costing*

Another example of applying Life Cycle Tools in Design is using Life Cycle Costing (LCC, see Sect. 2.3) in Design for Environment (DfE) product development. The intention is to identify whether there is any kind of business case for different design options when considering all costs, including those beyond the direct producer costs (for development, production, marketing, etc.) and, in particular, the costs of their consumers (cost of ownership including energy, tax, insurance, maintenance, etc.) as well as the costs of product disposal [51, 52]. This approach improves the competitiveness of products from the start, as it considers the costs of consumers as well as potential future liabilities or economic risks (cost-free take-back or loss of acceptance if the disposal is not solved/too costly/with a negative effect of residual values). The actual consumer costs will vary by market (tax, energy costs, etc.). The residual value of the product might also be included. Based on this, a business case might exist for design options that increase the product cost but reduce the cost of ownership or other cost risks by, for example, improving energy/auxiliaries use, durability, or serviceability. The basic requirement for such a DfE business cases (type 1) is [53]:

$$\begin{aligned} S(\text{DfE})_D + !S(\text{DfE})_P + S(\text{DfE})_U + S(\text{DfE})_R \\ > C(\text{DfE})_D + C(\text{DfE})_P + C(\text{DfE})_U + C(\text{DfE})_R \end{aligned} \tag{4.1}$$

$S(\text{DfE})_D$ = Savings in product development due to the DfE action
$S(\text{DfE})_P$ = Savings in production due to the DfE action
$S(\text{DfE})_U$ = Savings in use due to the DfE action (including energy, auxiliaries, taxation, maintenance, insurance, residual value, etc.)
$S(\text{DfE})_R$ = Savings in recycling/disposal due to the DfE action
$C(\text{DfE})_D$ = Added Cost of DfE action in design
$C(\text{DfE})_P$ = Added Cost of DfE action in production
$C(\text{DfE})_U$ = Added Cost of DfE action in use (e.g. heavier but recycled materials in vehicles)
$C(\text{DfE})_R$ = Added Cost of DfE action in recycling (e.g. by composites instead of mono-materials)

This approach is a first step beyond the classical calculation of business cases:

$$S(\text{DfE})_P + S(\text{DfE})_P > C(\text{DfE})_D + C(\text{DfE})_P \tag{4.2}$$

An environmental business case can be made for designs where the resulting Life Cycle savings are higher than potential added costs and where the design offers an environmentally superior performance when compared to alternatives. Of course, there are situations where the overall costs increase while the environmental benefits increase as well, i.e. where trade-offs exist. Here, the environmental business case can be made by considering environmental efficiencies. This means that the alternative is chosen that offers the highest environmental improvement per additional cost unit. Environmental performance can be evaluated by using the tool of Life Cycle Assessment and is expressed as the result of the Life Cycle Impact Assessment (see Sect. 2.2.3). This makes sense where the fundamental decision/assumption is that an environmental improvement has to be achieved and the only question is how this can be achieved. This is obviously a different approach than the one represented in Eq. 4.1 where DfE actions are only supported where overall cost savings can be demonstrated.

The environmental business case type 2 is a situation where the Life Cycle Costs of alternatives are equal or higher than the existing design (Eq. 4.3) while the improved environmental performance per unit of additional cost is higher than in the case of other alternatives (Eq. 4.4).

An environmental business case type 1 has higher life cycle savings than costs. Type 2 is the alternative with the best environmental improvement per additional unit of life cycle cost.

The alternatives considered could be material changes, technology alternatives for the same product, component, etc. or alternatives for the product itself (like service instead of product approaches). Obviously, it is important that an environmental improvement can be demonstrated (Eq. 4.5).

$$LCC_{\text{Alternatives1}} \geq LCC_{\text{Base}} \tag{4.3}$$

$$\begin{aligned} EE_1(x) &= \frac{LCIA(x)_{\text{Alternative1}} - LCIA(x)_{\text{Base}}}{LCC_{\text{Alternative1}} - LCC_{\text{Base}}} \\ &> -\frac{LCIA(x)_{\text{Alternative2}} - LCIA(x)_{\text{Base}}}{LCC_{\text{Alternative2}} - LCC_{\text{Base}}} = EE_2(x) \end{aligned} \tag{4.4}$$

$$EE_1(x) > 0 \tag{4.5}$$

$EE(x)$ = Environmental efficiency of Life Cycle Impact Assessment Category x
$LCIA(x)$ = Life Cycle Impact Assessment result of design for category x
LCC = Life Cycle Cost result of design

References

1. B.N. Davidsson, *Modified Product Quality Tools for Improved Environmental Design in Small and Medium Sized Enterprises, Lund* (The International Institute for Industrial Environmental Economics at Lund University (IIIEE), Sweden, 1998)
2. D.L. Meadows, D. Meadows, J. Randers, W.W. Behrens, *The Limits to Growth—Report for The Club of Rome* (Potomac Associates—Universe Books, New York, 1972)
3. T.U. Delft, Compatibility of materials for recycling. Designing Electronics for Recycling in a Circular Economy, Delft
4. K. Kazmierczak, S.E. Mathiassen, M. Forsman, J. Winkel, An Integrated Analysis of Ergonomics and Time Consumption in Swedish 'Craft-Type' Car Disassembly. Appl. Ergon. **36**(3), 263–273 (May 2005)
5. J.-C. Alonso, W.-P. Schmidt, J. Fontanilles, *Recyclability Improvement of the Vehicle Electrical & Electronic System* (Barcelona, Spain, 2004).
6. K. Stefan, Life Cycle Assessment of the Volkswagen-Sicon Process, 2005.
7. T. Nürrenbach, Recovery of Plastic Parts from ELV's—Evaluation of Environmental Impacts According to the LCA Method and Economic Analysis, Freising, 2003.
8. W. Jenseit, H. Stahl, V. Wollny, W. Rolf, Recovery Options for Plastic Parts from End- of-Life Vehicles: An Eco-Efficiency Assessment, Darmstadt, 2003.
9. M.B.G. Castro, J.A.M. Remmerswaal, M.A. Reuter, Life Cycle Impact Assessment of the Average Passenger Vehicle in the Netherlands. Int. J. LCA **8**(5), 297–304 (2003)
10. W.-P. Schmidt, E. Dahlqvist, M. Finkbeiner, S. Krinke, S. Lazzari, D. Oschmann, S. Pichon, C. Thiel, Life Cycle Assessment of Lightweight and End-of-Life Scenarios for Generic Compact Class Passenger Vehicles. Int. J. LCA **9**(6), 405–416 (2004)
11. M.A. Reuter, A. van Schaik, O. Ignatenko, G.J. de Haan, Fundamental Limits for the Recycling of End-of-Life Vehicles. Miner. Eng. **19**(5), 433–449 (2006)
12. E. Pestel, *Beyond the Limits to Growth—A Report to the Club of Rome* (Universe Books, 1989)
13. D. Meadows, J. Randers, D.L. Meadows, *Limits to Growth: The 30-Year Update* (Earthscan, London, 2004)
14. D. Huberts, Fuel Cells Meet Big Business. *The Economist,* Quote Page 59 (24 July 1999)
15. W.-P. Schmidt, *Ökologische Grenzkosten der Kreislaufwirtschaft* (TU Berlin, Berlin, 1997)
16. European Comission, Study on the Critical Raw Materials for the EU 2023—Final Report, Brussels
17. N. Tazi, M. Orefice, C. Marmy, Y. Baron, M. Ljunggren, P. Wäger, F. Mathieux, *Initial Analysis of Selected Measures to Improve the Circularity of Critical Raw Materials and Other Materials in Vehicles* (Publications Office of the European Union, Luxembourg, 2023)
18. The International Dismantling Information System, https://www.idis2.com/
19. P. Morseletto, Targets for a Circular Economy. Resour. Conserv. Recycl. **153**(104553) (2020)
20. W. McDonough, M. Braungart, *Cradle to Cradle: Remaking the Way We Make Things* (North Point Press, New York, 2002)
21. E.B. Kennedy, Biomimicry: Design by Analogy to Biology. Res. Technol. Manage. **60**(6), 51–56
22. ISO 59004, "Circular Economy – Terminology, Principles and Guidance for Implementation," International Standards Organization, 2021.
23. ISO 59020, Circular Economy—Measuring and Assessing Circularity, International Standards Organization, 2021
24. S. Jones, T. Krieger, B. Coleman, S. Iliefski-Janols, R. Veenendal, P. Stoltz, *Circularity Indicators—An Approach to Measuring Circularity—Methodology* (A. G. Ellen MacArthur Foundation, Ed., 2015)
25. R. Wakkary, A. Desjardins, H. Lin, L. Maestri, Material Design for Sustainability: An Emergent Process Model. Des. Stud. **47**, 111–139 (2016)
26. F. Quella, W.-P. Schmidt, Integrating Environmental Aspects into Product Design and Development: The New ISO TR 14062—Part 2: Contents and Practical Solutions. *Gate to EHS: Life Cycle Management—Design for Environment* (17 March 2003), p. 7

27. G. Rebitzer, W.-P. Schmidt, Design for Environment in the Automotive Sector with the Materials Selection Tool euroMat. *Gate to EHS: Life Cycle Management—Design for Environment* (17 March 2003), pp. 1–4
28. SAE, Potential Failure Mode and Effects Analysis in Design (Design FMEA) and Potential Failure Mode and Effects in Manufacturing and Assembly Processes (Process FMEA) reference manual, July 1994
29. E. Östebo, *Environmental Impact Analysis, E-FMEA* (Volvo, Gothenbourg, 1999)
30. W.-P. Schmidt, Umwelt-Fehlermöglichkeiten- und Einfluss-Analyse, in *Umweltgerechte Produktentwicklung – Ein Leitfaden für Entwicklung und Konstruktion* (Wien Zürich, Beuth Verlag, Berlin, 2001), p.12
31. A. Neligan, C. Schleicher, B. Engels, T. Kroke, *Digitaler Produktpass – Enabler der Circular Economy. Relevanz und Umsetzbarkeit durch Unternehmen* (Institut der deutschen Wirtschaft Köln e. V. , Berlin/Köln, 22 September 2023)
32. H. Kunst, Ökologische Optimierung von Substitutionsentscheidungen, Dissertation.de, Berlin, 2003, p. 178
33. M. Bovea, V. Pérez-Belis, A Taxonomy of Ecodesign Tools for Integrating Environmental Requirements into the Product Design Process. J. Clean. Prod. **20**, 60–71 (2012)
34. S. Jagtap, A. Larsson, V. Hiort, E. Olander, A. Warell, P. Khadilkar, How Design Process for the Base of the Pyramid Differs from That for the Top of the Pyramid. Des. Stud. **35**(5), 527–558 (2014)
35. V. Rizos, J. Bryhn, M. Alessi, A. Campmas, A. Zarra, *Identifying the Impact of the Circular Economy on the Fast-Moving Consumer Goods Industry: opportunities and Challenges for Businesses, Workers and Consumers—Mobile Phones as an Example* (C. f. E. P. Studies, Ed., 2019)
36. C. Cawley, 10 Creative Ways to Use Your Old Smartphone, 2021
37. S. Suppipat, A.H. Hu, A Scoping Review of Design for Circularity in the Electrical and Electronics Industry. Resources, Conservation & Recycling Advances **13**, 2667–3789 (2022)
38. J. Chapman, *Emotionally Durable Design* (Routledge, London, 2005)
39. Renault, ReFactory, Renault (25 November 2020), https://www.renaultgroup.com/en/news-on-air/news/re-factory-the-flins-site-enters-the-circle-of-the-circular-economy/. Accessed 8 April 2023
40. UNEP, ed., E. Manzini, C. Vezzoli, *Product Service Systems and Sustainability—Opportunities for Sustainable Solutions* (UNEP, Paris, France, 2000)
41. K.F. Gustafsson, S.A. Harild, D.C.A. Pigosso, Design for Sustainable Behaviour in Product/ Service Systems—A Systematic Review. *Proceedings of the Design Society* (2021), pp. 1033–1042
42. F. Ceschin, I. Gaziulusoy, Evolution of Design for Sustainability: From Product Design to Design for System Innovations and Transitions. Des. Stud. **47**, 118–163 (2016)
43. Leichtbauwelt—Inspiration für Ihren Fortschritt im Leichtbau, https://www.leichtbauwelt.de/ 5-beispiele-fuer-bionik-im-leichtbau/. Accessed 3 April 2024
44. R.J. Tapper, M.L. Longana, A. Norton, K.D. Potter, I. Hamerton, An Evaluation of Life Cycle Assessment and Its Application to the Closed-Loop Recycling of Carbon Fibre Reinforced Polymers. Compos. Part B Eng. **184**(107665) (2020)
45. W.-P. Schmidt, C. Duranceau, J. Sullivan, Sustainable Materials in Automotive Applications, in *Proceedings of the 2002 Environmental Sustainability Conference and Exhibition* (2002), pp. 393–398
46. ISO 14021:2016, Environmental Labels and Declarations—Self-Declared Environmental Claims (Type II Environmental Labelling), Geneva, 2016
47. T.U. Berlin, T.U. Cottbus, C.A.U Gmbh, IKV, IPT, ICT, Systematische Auswahlkriterien für die Entwicklung von Verbundwerkstoffen unter Beachtung ökologischer Erfordernisse (euroMat), Berlin, 1995
48. I. Magnet, Magnetismus - Dauermagnete - Werkstoffe und Systeme, Berlin
49. W.-P. Schmidt, F. Butt, Life Cycle Tools within Ford of Europe's Product Sustainability Index. Case Study Ford S-MAX & Ford Galaxy. Int. J. LCA **11**(5), 315–322 (2006)

50. Ford, Product Sustainability Index, Ford of Europe, Cologne, 2006
51. K. Ehrlenspiel, H. Meerkamm, *Integrierte Produktentwicklung – Methoden für Prozeßorganisation, Produkterstellung und Konstruktion* (Hanser, München, 1995)
52. G. Rebitzer, Modul Kosten, in *Eco-Design: Effiziente Entwicklung nachhaltiger Produkte mit euroMat* (Springer, Berlin, Heidelberg, 2000), p. 103–112
53. W.-P. Schmidt, Life Cycle Costing as Part of Design for Environment – Environmental Business Cases. Int. J. Life Cycle Assess. **8**(3), 167–174 (2003)

Chapter 5
Sustainable Supply Chain Management, Purchasing and Logistics

Abstract Purchasing/procurement by an organization can play a fundamental and powerful role in achieving more sustainable life cycle performances, allowing compliance with UN Sustainable Development Goals. Tenders and (de-)selection of suppliers send clear market demand signals which impact upstream life cycle actors and potentially provide them with additional reasons for the required business transformation—especially if this is complemented by a strategic partner approach and/or if the organization has a purchasing volume of interest for the upstream stakeholders. In addition, the evolving regulatory framework for sustainable supply chain management is described. Finally, assessment of the sustainability of purchasing decisions must also consider logistics. With a share of 60%, transportation is the biggest consumer of oil worldwide and accounts for around a quarter of global Greenhouse Gas Emissions from energy. This chapter describes opportunities to optimize logistical models to achieve the lowest sustainability footprint possible.

Keywords Due diligence · Sustainable assessment questionnaires · Regulation · Logistics · Transport modes

The principles of sustainable supply chain management and purchasing apply to situations where an organization is the customer of upstream life cycle stakeholders in a business-to-business (B2B) relationship (Fig. 2.1). While similarities to sustainable end-customers exist, (Chap. 7) there are also specific B2B opportunities and challenges that are described in this chapter.

The purchasing/procurement function of an organization can play a central and powerful role in the sustainability life cycle performance in terms of compliance with UN Sustainable Development Goals. The tenders and (de-)selection of suppliers send clear market demand signals which impact upstream life cycle actors and potentially provide them with additional reasons for the required business transformation—especially if this is complemented by a strategic partner approach and/or if the organization has a purchasing volume of interest for the upstream stakeholders. In addition, the regulatory framework is more developed for B2B sustainable purchasing than for end-customers (Table 5.1).

© The Author(s), under exclusive license to Springer Nature Switzerland AG 2024
W. Schmidt, *Solutions For Sustainability Challenges*,
https://doi.org/10.1007/978-3-031-63624-0_5

Table 5.1 Practices available in the upstream supply chain to implement sustainable business models ([1] modified)

Approach	Example
Selection/ evaluation of suppliers	• Purchase of more sustainable materials • Sustainability policy and documentation requirements for suppliers/contractors • Selection of suppliers/subcontractors based on sustainability criteria • Scouting for new suppliers with specific skills in transformation/sustainability or more sustainable products, lower carbon footprint etc • Simplification of the supply chain • Consideration of sustainability standards and certifications
Objective setting with suppliers	• Agreement on clear sustainability objectives for the supplied service/product
Monitoring of suppliers	• Analysis of incoming materials based on sustainability criteria • Monitoring of the sustainability performance of suppliers/subcontractors • Traceability and actions to achieve full visibility based on specific data for the 1st, 2nd and subsequent tiers of supply (e.g. blockchain)
Collaboration with suppliers	• Knowledge sharing about sustainability • Sustainable packaging, product improvement • Improving sustainability logistics • Recovery of production waste and waste management • Recovery of the product at the end of its life cycle • Joint review meetings on sustainability KPIs
Integration with suppliers/ business customers	• Joint sustainability projects • Joint investments in more sustainable production processes, machines, energy generation, etc • Strategic sustainability partnerships • Sharing sustainability goals • IT integration to verify sustainability performance within the supply chain (e.g. Catena-X [2], IMDS [3]) • Circular economy between partners

Sustainable Purchasing sends clear market signals to suppliers and supports a wider transformation in industry especially if implemented in a strategic, cooperative approach.

5.1 Supply Chain Due Diligence

The globalization and complexity of most product's supply chains run the risk of hidden sustainability issues that potentially negatively impact the reputation and acceptance of products. Examples include the

- Rana Plaza factory collapse in Bangladesh in 2013 with more than 1000 killed workers that demonstrated poor working conditions and safety practices in the garment industry [4].
- Child and forced labor in the cocoa industry, particularly in West Africa but also closer to Europe where child labor in Türkiye is used to collect hazelnuts for hazelnut nougat cremes [5].
- Deforestation and illegal logging, particularly in the Amazon rainforest that has led and is leading to environmental degradation, loss of biodiversity, and displacement of indigenous communities [6]. Similar issues are seen, for example, in Romania.
- The use of minerals from conflict zones in electronics manufacturing. Particularly in the Democratic Republic of Congo, this is fueling the ongoing violence and human rights abuses [7].

The Drive Sustainability initiative of the automotive industry is an example where a sector has tried to establish a collective management of knowledge to provide a basis for individual companies to act with due diligence (Fig. 5.1).

Due diligence is a "process through which organizations proactively identify, assess, prevent, mitigate and account for how they address their actual and potential adverse impacts as an integral part of decision-making and risk management" [9]

The OECD suggested a five-step approach in 2016 [7]:

1. Establish strong company management systems.
2. Identify and assess risks in the supply chain.
3. Design and implement a strategy to respond to the identified risks.
4. Carry out independent third-party audits of suppliers (smelter/refiners) due diligence practices.
5. Report annually on supply chain due diligence.

Associations have developed standard questionnaires for the assessment of risks in the supply chain. This avoids suppliers having to answer similar and overlapping questions from multiple business customers and to allow association members to

Fig. 5.1 Drive sustainability knowledge management [8] © CSR Europe

identify their greatest risks. Examples include the Responsible Business Alliance (RBA) Self-Assessment Questionnaire (SAQ) that contains 80 questions related to ethics, labor, environment, and health & safety. Example questions include "Who is in possession of workers' original documents like passports, ID cards?", "Is water recycled at your facility", "Do you communicate RBA Code of Conduct requirements to your suppliers?". Based on responses and additional information a risk score is calculated for each category. These scores are combined with other risk factors, such as the general risk associated with the country. If the resulting risk score is too high, a detailed audit as well as a Corrective Action Plan should be conducted. The Corrective Action Plan details the SAQ findings to be addressed, the corrective action to be taken, the responsible person/organization, the target completion date, and evidence for the correction as well as actions to prevent long-term recurrence.

In September 2023, Drive Sustainability issued a first report that analyzed how their Sustainability Assessment Questionnaires (SAQ) have been answered since their introduction in 2014. Around 55,000 suppliers completed around 114,000 SAQs in this time period. Their findings which were based on the latest SAQs between 2021 and 2023 conclude that SAQ performance [10]:

- Is significantly impacted by company's employee headcount and sector while company location is only relevant for small and medium-sized enterprise's (SMEs),
- There was improvement over time (12% for larger companies, 20% for SMEs). This might have been the result of the SAQs and/or other factors.

The identified weaker SAQ performance of SMEs with less than 250 employees (around 70% of analyzed SAQ dataset) is most prominent in the areas of human rights and business ethics where less employee training occurs.

Following the SAQ screening, actual audits are needed for low-scoring cases. For example, audits are needed to check for forced labor or modern slavery. Such practices also happen in OECD countries and need to be carefully detected by looking at signs of missing consent or menace of penalty (see Sect. 1.2). These include:

- Recruitment processes linked to high fees.
- Limited ability and openness of communication between workers and supervisors.
- Limited freedom of movement for workers (e.g. by checking contracts conditions, where passports or other properties are stored, rules limiting usage of toilets or ability to leave the working shift).
- Physical or sexual abuse including signs of injury, having workers in locations with unacceptable temperatures, turning off fans, requiring push-ups,
- Excessive unpaid overtime with threats of job termination if workers refuse.
- Withheld wages, low net wage, delayed payment.

Due Diligence can be based on screening questionaries, targeted audits, or validated corrective actions. Human rights and business ethics are more often an issue for SMEs.

Several regulations are in place to address sustainable supply chain due diligence und human rights including:

- US Uyghur Forced Labor Prevention Act implemented in June 2022. This act bans the importation of goods mined, produced, or manufactured wholly or in part with forced labor from the Xinjiang region of China into the US. The act presumes that imported goods from the Xinjiang region are made with forced labor and therefore cannot enter the US. Importing companies are responsible to prove goods are not manufactured with forced labor to release detained goods (within 30 days of detention notice).
- Several countries and regions have established formal Forced Labor Bans that apply globally. Examples include Mexico (effective May 18th, 2023) banning imports made with force labor as well as Canada's Fighting Against Forced Labour and Child Labour in Supply Chains Act (effective May 11th, 2023) that demands a detailed due diligence process including risk identification, remedial actions taken, training, effectiveness assessment, and annual reporting. Similar acts are the UK Modern Slavery Act, the California Transparency Act, and the Australian Commonwealth Modern Slavery Act. The Australian Act is likely to extend its rules from 2018. For example, entities must now have due diligence management systems, measures, and administrative guidance in place.
- National due diligence regulation reports are mandated, for example, by the Swiss Ordinance on Due Diligence and Transparency (focus on conflict minerals and

child labor), the Swedish Due Diligence regulation, the Norwegian Transparency Act (focus on fundamental human rights including decent working conditions) and on a US state level the Californian Transparency in Supply Chain Act.

- The German Supply Chain Due Diligence Act entered into force on August 18, 2023 and requires companies to establish various risk management procedures to safeguard human rights and the environment. This includes an annual risk analysis and preventative measures for the organizations' own global facilities and business areas including subsidiaries where the parent company exerts a decisive influence as well as for its direct suppliers and indirect suppliers where "substantiated knowledge" suggests potential violations. The regulation includes a grievance mechanism, remedial actions, policy statements on human rights strategy policies and processes, yearly public reports on due diligence procedures, risks identified, and measures taken. In addition, a human rights ombudsperson is required. Actions should focus on high-risk areas where the organization has a high degree of influence (share of business) and where there is a causal contribution first.
- The EU Corporate Sustainability Due Diligence Directive (CS3D) partly expands on the German Supply Chain Due Diligence Act e.g. by covering the entire value chain based on a risk assessment approach and by emphasizing due diligence obligations also in relation to adverse environmental impacts. CS3D considers the company level, business model, geographic region, product, and service as well as sectoral risk factors as the basis for the risk assessment to decide on what suppliers to focus on. An important additional aspect is that each company is required to report whether its business model is aligned with the goal of limiting climate change to 1.5 °C above preindustrial levels. Actions are expected for high-risk activities that cause, contributed to, are linked to or can be leveraged against high impacts on sustainability or human rights.
- On an EU level, due diligence requirements are also referenced in other regulations like the EU Battery Regulation that came into force on August 18, 2023.
- The EU Regulation on Deforestation-free Products adds a biodiversity aspect into due diligence requirements and came into force on June 29, 2023. It requires due diligence (risk assessment and mitigation) for a few explicitly mentioned food products like cattle, coffee, cocoa, palm oil, and soya but also non-food products like rubber and rubber-containing products like tires or timber and wood containing products like books, newspaper, and wooden furniture. These products need to be free of deforestation which is defined as "the conversion of forest to agricultural use, whether human-induced or not" [11]. The UK has announced its intention to regulate the use of so-called "Forest Risk Commodities" (FRC) in commercial as well.
- The EU Corporate Sustainability Reporting Directive is mainly a disclosure and reporting requirement (see also Sect. 7.7.1) (Table 5.2).

Table 5.2 Supply chain regulations

Regulation	Examples	Scope (Area of protection, supply chain)	Requirement examples
Supply chain regulation targeted on one region/issue	• US Uyghur Forced Labor Prevention • US Dodd Frank Act	• Forced Labor only in Xinjiang region (all imports) • 4 minerals from the Democratic Republic of Congo	• Within 30 days of detention notice prove goods are not manufactured with forced labor to release detained goods • Due diligence reporting to avoid financing violence
Ban of products linked to forced or child labor, modern slavery	Regulations in Australia, Canada, California, Mexico, UK, etc	Forced or child labor imported from all regions	Canada: detailed due diligence process including risk identification, remedial actions taken, training, effectiveness assessment, annual reporting
Ban of products compromising biodiversity	EU Regulation on Deforestation-free Products	Certain products/products containing e.g. cattle, coffee, cocoa, palm oil, soya, rubber, timber	Risk assessment and mitigation
Broad sustainability Due Diligence	• German Act on Corporate Due Diligence for the Prevention of Human Rights Violations in supply chains • EU CS3D	• Human rights, environmental risks from substances and waste. Own business, direct supplier, and where substantiated knowledge exists also indirect suppliers • EU similar but all supply chain based on risk assessment	• Grievance mechanism, remedial actions, human rights strategy policies and processes, yearly public reports on due diligence procedures, risks identified, and measures taken. In addition, a human rights ombudsperson is required. Annual risk analysis and preventative measures for own operations • Report whether the business model aligns with the goal of limiting climate change to 1.5 °C

© Wulf-Peter Schmidt, all rights reserved

Due diligence regulations range in scope from limited issues (forced labor in one region) to broad environmental and human right issues in the global supply chain based on risk assessments.

5.2 Sustainable Procurement

ISO 20400:2017 (Sustainable procurement—Guidelines) provides guidance for companies (private, public, all sizes) on how to integrate sustainability into procurement practices. It defines sustainable procurement as "the process of making purchasing decisions that meet an organization's needs for goods and services in a way that benefits not only the organization but society as a whole, while minimizing its impact on the environment. This is achieved by ensuring that the working conditions of its suppliers' employees are decent, the products or services purchased are sustainable, where possible, and that socio-economic issues, such as inequality and poverty, are addressed" [9]. This refers specifically to the procurement process that has "the most positive environmental, social and economic impacts possible across the entire life cycle and that strives to minimize adverse impacts".

ISO 20400:2017 complements ISO 26000:2010, Guidance on social responsibility, by considering specifically the function of the purchase. This guidance document is not a standard that organizations can be certified for. It is based on key principles including:

- Accountability—for the impacts of the organization's supply chain as well.
- Transparency—in its procurement decisions to enable dialogue and collaboration with stakeholders including suppliers.
- Ethical behavior, avoidance of bias and prejudice in all procurement decisions.
- Respect for the rule of law, international norms of behavior, human rights stakeholder interests.
- Analysis of all costs along the life cycle (with reference to Life Cycle Costing; see Sect. 2.3).
- Integration of sustainability in all procurement processes.
- Focus on needs—analyze real demand, buy only what is really needed and seek more sustainable alternatives.
- Continuous improvement.

The seven core subjects of sustainable procurement are the following [9]:

- Governance: decision-making processes and structures;

- Human rights: due diligence, human rights risk situations, avoidance of complicity, resolving grievances, discrimination and vulnerable groups, civil and political rights, economic, social and cultural rights, fundamental principles and rights at work;
- Labor practices: employment and employment relationships, conditions of work and social protection, social dialogue, health and safety at work, human development and training in the workplace;
- Environment: prevention of pollution, sustainable resource use, climate change mitigation and adaptation, protection of the environment, biodiversity, and restoration of natural habitats;
- Fair operating practices: anti-corruption, responsible political involvement, fair competition, promoting sustainability in the value chain, respect for property rights;
- Consumer issues: fair marketing, factual and unbiased information, fair contractual practices, protecting consumers' health and safety, sustainable consumption, consumer service and support, complaint and dispute resolution, consumer data protection and privacy, access to essential services, education and awareness;
- Community involvement and development: community involvement, education and culture, employment creation and skills development, technology development and access, wealth and income creation, health, and social investment.

> Sustainable procurement is based on good governance, fair operating practices, ethical behaviors, accountability, a consideration of all life cycle sustainability aspects and a focus on real needs.

Beyond the ISO requirements, it is key to have a clear selection and evaluation approach for suppliers as well as clear sustainability KPIs and agreements. Sustainable procurement must also consider the life cycle implications of the procurement decision like the impact on transport distances/logistics.

Typically, larger companies require their suppliers to comply with all regulations, human right obligations, follow ethical behavior, and also add minimum environmental requirements (like ISO 14001). They also expect due diligence to be applied to their sub-supplier chain. The automotive industry also has sector-wide supplier reporting tools like the International Material Data System [3] where the substances and materials of the products are reported along the whole supply chain.

5.3 Sustainable Supply Chain Network Partnership

The Business-to-Business (B2B) relationship between suppliers and manufacturers who purchase pre-products, systems, and materials is key to improve results from a sustainability perspective. Seeing suppliers as an outsourced, cost-optimized delivery system of a commodity that fits the manufacturer's production requirements is a missed opportunity. The life cycle perspective should be used to:

- Create a network of companies that complement each other and establish material and recycling flows between all of them (circular economy).
- Offer a platform to exchange best practices between all companies (e.g. Ford's Partnership for a Cleaner Environment, PACE [12]).
- Be open to innovation and ideas for product and service improvements.
- Create new products in close collaboration with customers and supply chain partners.

5.4 Sustainable Goods Transportation

With a share of 60%, transportation is the biggest consumer of oil worldwide and accounts for around a quarter of the global Greenhouse Gas Emissions from energy [13].

Logistics is the "process of planning, implementing, and controlling the efficient, effective flow and storage of goods, services, and related information from point of origin to point of consumption for the purpose of conforming to customer requirements". (Definition of the American Council of Logistics Management). Considerations in logistics must be broadened beyond cost and time efficiency to include sustainability. This includes minimizing any environmental impacts (ideally to close to zero-emission transportation) while also respecting social implications like accessibility, movement and activities, health-related, financial-related, and community-related impacts as well as spatial, temporal, and socio-demographic distributional effects of transportation.

Transportation is required for and between all physical processes (Fig. 2.1). There is in-bound (incoming including returning), out-bound (sending processed products out), and on-site transportation. Transportation is particularly impacted by:

- Locations/distances (defined by own production sites as well as by purchasing decisions that need to consider sustainable logistics).
- Type and packaging of transported goods (weight, volume, conditions like temperature control).

- Transport mode: by sea, land, air/by constant flows (e.g. in pipelines) or discrete batches (like in trucks, ships, planes, trains).
- Abilities and constraints of infrastructure and transport modes (e.g. traffic jams/ capacities, up-time, up vs. down the river).
- Logistical models (just-in-time, hub distributions, storage, utilization grade of transport mode, return trips, etc.).

Sustainable logistics needs to optimize all of these aspects, i.e. reduce distances (task together with the purchasing department), optimize packaging, select transport modes with the lowest impact while also considering real-world impacts/constraints and optimizing logistical models towards the lowest sustainability footprint possible.

Sustainable transport optimizes distances, packaging, transport mode & models. Avoidance, cargo bike, pipelines, railways, vehicles, and airfreight have consecutively higher carbon footprints.

Usually, LCAs show that the impact of transportation on the overall life cycle is small even if linked to long distances unless air freight is involved. The reason for the relatively low environmental impact is that from an individual product perspective the emissions of a truck, ship, or train are divided by the number of goods transported (see Sect. 2.1.4) (Table 5.3).

The cost efficiency of these options varies significantly and may depend on CO_2 pricing (e.g. lorry toll systems). Regulatory costs from CO_2 pricing are currently relatively high (a 200–300 €/ton CO_2 reduction by HVO trucks or 700–800 €/ton CO_2 reduction by HVO shipping might be achieved) although cost parity is suggested for some options like electrified ocean shipping but only when compared to airfreight [15].

From an environmental LCA perspective, [18] a Battery Electric Lorry (12 tons/ 40 tons) is the best road vehicle option from an energy and global warming potential perspective followed by Fuel Cell Electric vehicles. From a social perspective, the working conditions of drivers are of high concern.

Table 5.3 Goods transport modes—opportunities, challenges and trends

	Sustainability opportunities	Sustainability challenges	Examples, trends
Avoidance	Lowest/no resources and emissions	Acceptance, local shortages	Local 3D printing instead of physical transport
Cargo bikes	Lowest environmental impact, last mile delivery, no traffic jams	Only short distances and small cargo, biker rights	Electrification, inclusion in logistic hub strategies
Pipelines	Low transport energy, low OPEX, lower footprint, requires less stock/storage due to constant inflow replenishment	Leakage (e.g. gas pipelines), vulnerability, investment, only for high and constant volume flows, not flexible, storage at the input side	Underground freight pipelines with electric-powered freight capsules [14] (not implemented yet)
Ships	Low energy demands due to big loading capacity	Slow, emissions, limited to seaside and shippable rivers, time lost by modal changes, working conditions	Electrified ocean shipping with battery swapping [15] or fast hydrogen-powered hydro-foiling cargo ship to replace air freight [16], HVO or LNG ships
Railway	Low energy demand, capacity, and speed can be competitive if infrastructure is sufficient	Limited flexibility, time lost by modal changes unless facilities have a direct rail connection	Full electrification based on renewable electricity, separation of goods and people infrastructure to increase reliability
Vehicles	Flexibility, door-to-door capability over long distance	Traffic jams, emissions, footprint of infrastructure, Working conditions	Zero-emission vehicles (BEV, Fuel Cell EV). Hydrogenated Vegetable Oil (HVO) drop-in fuel
Airfreight	Fastest on long distance, limited infrastructure	Highest emissions/energy need, time lost by modal changes	Low-carbon fuel (so-called "Sustainable Aviation Fuels") [17]

© Wulf-Peter Schmidt, all rights reserved

References

1. L. Macchion, A.C. Toscani, A. Vinelli, Sustainable Business Models of Small and Medium-Sized Enterprises and the Relationships to be Established Within the Supply Chain to Support These Models. Corp. Soc. Responsib. Environ. Manag. **30**(2), 563–573 (2023)
2. Catena-X—Your Automotive Network. https://catena-x.net/en/
3. DXC Technologies, Material Data System. https://public.mdsystem.com/en/web/imds-public-pages
4. P. Koenig, S. Poncet, The Effects of the Rana Plaza Collapse on the Sourcing Choices of French Importers. J. Int. Econ. **137**, 103576 (2022)
5. Fair Labor Association, Hazelnuts in Turkey. Report Prepared by the Fair Labor Association (FLA) (Washington, 2017)
6. A. Dascăalu, J. Catalão, A. Navarro, Detecting Deforestation Using Logistic Analysis and Sentinel-1 Multitemporal Backscatter Data. Remote. Sens. **15**, 290 (2023)

7. OECD, *OECD Due Diligence Guidance for Responsible Supply Chains of Minerals from Conflict-Affected and High-Risk Areas*, 3rd edn. (OECD Publishing, Paris, 2016)
8. Drive Sustainability Develops Raw Material Outlook (2021). https://www.drivesustainability.org/mediaroom/drive-sustainability-develops-raw-material-outlook/. Accessed 26 July 2024
9. ISO (20400:2017), *Sustainable Procurement—Guidance* (International Standard Organization, Geneva, 2017)
10. S. A. Drive Sustainability, Supply Chain Due Diligence—How is the Automotive Industry Improving Sustainability Performance? (2023)
11. European Union, Regulation (EU) 2023/1115 on the Making Available on the Union Market and the Export from the Union of Certain Commodities and Products Associated with Deforestation and Forest Degradation and Repealing Regulation (EU) No 995/2010. (Brussels 2023)
12. UN Global Combat, Partnership for a Cleaner Environment (PACE) (2017). https://ceowatermandate.org/resources/partnership-for-a-cleaner-environment-pace-2017/. Accessed 25 July 2024
13. Circle Economy, *The Circularity Gap Report 2023* (Circle Economy, Amsterdam, 2023)
14. CargoCap GmbH, CargoCap—Gütertransport im Ballungsraum. http://www.cargocap.de/. Accessed 29 April, 2024
15. Fleetzero, Fleetzero—Shipping Electrified. https://fleetzero.com/. Accessed 29 April 2024
16. Boundary Layer Technologies, ARGO—A Hydrogen-Powered Hydrofoiling Cargo Ship to Replace Air Freight. https://www.boundarylayer.tech/argo. Accessed 29 April 2024
17. P. Kurzawskaa, Overview of Sustainable Aviation Fuels Including Emission of Particulate Matter and Harmful Gaseous Exhaust Gas Compounds. Transp. Res. Procedia **59**, 38–45 (2021)
18. N. Hill, S. Amaral, S. Morgan-Price, T. Nokes, J. Bates, H. H. Helms, K. Biemann, N. Abdalla, J. Jöhrens, E. Cotton, L. German, A. Harris, S. Haye, C. Sim, Determining the Environmental Impacts of Conventional and Alternatively Fuelled Vehicles through LCA (2020)

Chapter 6
Sustainable Production

Abstract Production processes have both positive and negative impacts on the sustainability of a product and the wider organization. Positive impacts include the creation of economic value, providing employment, allowing development and fostering wealth as well as addressing customer needs. In addition, products can also have direct positive impacts by addressing UN Sustainable Development Goals. Ideally, an organization should center its business strategy and model around such products (see Chapter 3). However, negative impacts can co-exist. These impacts are often linked to materials or substances that require significant amounts of energy/water/chemicals. Negative impacts may result from processes that use lots of energy/water/chemicals/emissions or involve sub-standard working conditions. This chapter provides an overview of the generic engineering methods used in production and strategies to ensure the sustainability of these methods. Furthermore, opportunities and challenges of digitalization in sustainable production and other fields are analyzed. Examples covering those production processes and sectors with the biggest impacts (including EV batteries, steel, cement, energy as well as agricultural production) will be discussed.

Keywords Forming · Separating · Joining · Coating · Digitalization · Batteries · Green steel · Renewable energy · Hydrogen · Buildings · City planning · Carbon sequestration

The production process has both positive and negative impacts on the sustainability of a product and the organization itself. Positive impacts include the creation of economic value, providing employment, allowing development and fostering wealth as well as addressing customer needs. In addition, products can also have direct positive impacts by addressing UN Sustainable Development Goals. Examples of such products include wind turbines, water-saving technologies, medical products, etc. Ideally, the organization centers its business strategy and model around such products (see Chapter 3).

Even in these cases, however, negative impacts can occur. These impacts can often be linked to materials or substances that require significant amounts of energy/

© The Author(s), under exclusive license to Springer Nature Switzerland AG 2024

W. Schmidt, *Solutions For Sustainability Challenges*,
https://doi.org/10.1007/978-3-031-63624-0_6

water/chemicals to generate. They can be also linked to products where the processing of these materials or substances corresponds with the use of lots of energy/water/chemicals/emissions or where the working conditions are sub-standard.

This chapter provides an overview of the generic engineering methods used in production as well as strategies to ensure the sustainability of production. Based on this fundamental knowledge, examples will be provided, focusing on those production processes and sectors with the biggest impacts.

6.1 Engineering Methods and Sustainable Production Strategies

According to DIN 8580 [1], fundamental engineering methods include primary forming, forming, cutting or separating, joining, coating, and changing material properties (Fig. 6.1).

Primary forming processes are used to convert raw or scrap materials into a basic shaped and sized (intermediate) primary product. The primary forming processes create cohesion of material particles and form basic solid bodies. Primary forming or molding processes include:

- Liquid or pasty states like molten metals (e.g. casting of metals or ceramics, cement, forming of composite plastics, foaming).
- Plastic state like injection molding, pressing of plastic resin, calendaring.
- Powder or particle state like sand molding.
- Fiber state like pressing to hardboard or creation of paper.
- Gas, vapor or from ionized state like electrolytical or evaporative deposition.

The joining processes include compounding, filling, pressing, joining by master forming, joining by shaping, joining by welding, joining by soldering, gluing, and textile joining. The reverse cutting processes include dividing, machining in geometrically defined or undefined ways like chipping, removing, erosion, electro-etching, disassembling, and cleaning (abrasive blasting). Cutting processes can also directly influence the surface topography of products by changing the surface itself.

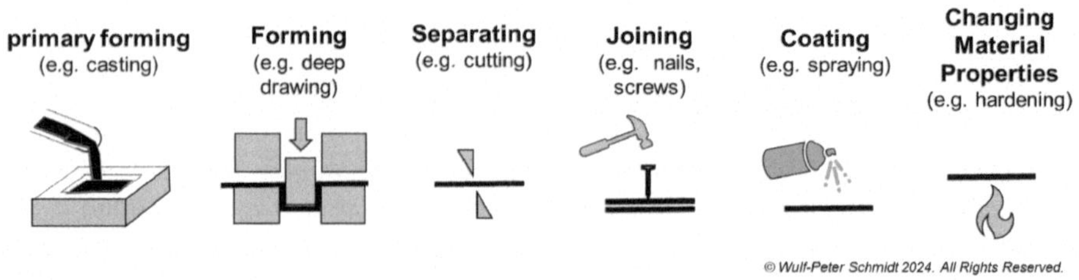

Fig. 6.1 Fundamental engineering method groups

The process of coating contains all surface treatment processes, which result in the addition of a coating onto a substrate. Coatings can be applied from liquid (hot dipping, dip coating, painting), solid (thermal spraying, electrostatic coating), gaseous, vapor (physical or chemical vapor deposition), or ionic (electro- and chemical plating) states of the coating material [2]. Coating can increase the durability and lifetime of products by increasing corrosion and wear-and-tear resistance as well as by reducing surface friction potentially leading to lower resource and energy consumption during product life, supporting their sustainability.

Processes that change material properties include those that influence the surface layer of products. For example, peening processes modify the hardness of the surface while heat treatment processes like induction hardening and plasma diffusion improve surfaces in various ways. Sintering processes also influence various material properties.

The sustainability impacts of these production processes depend especially on their energy consumption. Energy consumption is heavily impacted by the temperatures and pressure levels required for the different processes. Temperatures are typically the highest if metal or ceramic materials need to be molten in (primary) forming. Emissions can also be high for various other processes depending on the types of processed materials. The working conditions depend mainly on the level of automation and work organization.

Production is based on (primary) forming, cutting/separating, joining, coating, and changing material properties. Energy use, materials, automation, and work organization impact sustainability.

Sustainable production is ideally "non-polluting, conserving energy and natural resources, economically viable, safe and healthful for workers, communities and consumers, socially and creatively rewarding for all working people" [3]. The ideal production facility is architecturally blended into its environment with a minimal footprint, has no emissions but creates positive social benefits (Table 6.1).

6.2 Sustainable Digitalization

Smart data digitalization enables real-time information about production emissions, transportation, and storage allowing for faster reactions and optimization including the avoidance of empty return tours. Other applications include artificial intelligence predicting when or real-time monitoring systems alerting that a part or product needs to be replaced or be maintained. This avoids down-times or damages to products, machinery, and equipment. This ideally leads to longer lifetimes.

Table 6.1 Sustainable production strategies

Strategy	Example
Optimize build-up phase	• Reuse or repurpose existing production sites (e.g. Re-factory approach in Flins [4]) • Adjust to local conditions (e.g. architecture) • Involve, respect neighboring communities (e.g. early information; community workshop) • Minimize impacts of construction and logistics
Optimize Input (Quality and Quantity)	• Replace critical or harmful materials (e.g. coal by green hydrogen—see Sect. 6.4) • Less material, less energy used (e.g. internal recycling or waste heat recovery and system energy integration see Sect. 6.5) • Replace fossil fuels with cleaner energy sources (see Sect. 6.5)
Improve social effectiveness	• Optimal working conditions, health & safety (e.g. better ergonomics, working organization, automation)
Improve Efficiency—more and better product output with less input	• Replace batch processes by continuous production • Smart production • Integrate processes, delete process steps, etc. (see Sect. 6.2) • Increase durability/lifetime
Optimize Emissions (Quality and Quantity) and circularity	• Recover waste heat • Recover waste water • Recycle production waste
Optimize life cycle management	• Establish material and energy flows within and between industries (circular economy) • Maximize the lifetime of machinery and equipment (e.g. by predictive maintenance—see Sect. 6.2)
Optimize build-down phase	• Avoid contaminating sites • Build back to preindustrial status • Reuse buildings • Ensure social solutions for employees and region

Better information sharing along the life cycle—like the Digital Product Passport, the International Material Data System (IMDS), or the International Dismantling Information System (IDIS)—enable improved process like targeted and easy dismantling. The Digital Product Passport (DPP) is seen as an important enabler of the circular economy that has the potential to contribute to climate neutrality [5]. A DPP should include information relevant to life cycle stakeholders such as downstream producers, consumers, and end-of-life operators. This information includes the name of the manufacturer and precise product identifiers, materials used, technical and environmental properties along with repair, recycling, and disposal options. This can improve the quality of recycling products as shown in the example of EV battery recycling (see Sect. 8.5.2). However, it is clear that information itself is insufficient and requires more sustainable designs. Life cycle stakeholders should consider this information and act based on it in such a way that improves the overall sustainability performance. The more complicated, unstandardized, and irrelevant information is, the lower the likelihood of a positive impact. DPPs are, for example, required by the EU Eco-Design Directive.

Digital market places like Waste2Profit, Waste-Outlet, Excess Materials Exchange, Cirplus, and ScrapAd help repurpose waste and by-products of industry by sharing material-related data in a digital marketplace to help other businesses to include these materials as resources. Therefore, digitalization enables a circular economy.

Apps and other digitalized tools can improve public education and information sharing on sustainability (e.g. carbon footprint of products in shops). Apps can also be used to foster more sustainable behaviors by gamification (see sect. 4.4).

Real-time availability of data and improved data sharing along the life cycle are important tools that enable & encourage life cycle stakeholders to reduce the impacts of their products.

Digitalization has a huge potential to reduce environmental impacts by avoiding the physical transportation of humans for work etc. while also having—depending on personal preferences positive or negative –social implications. The overall sustainability impact depends on how digitalization opportunities are used by organizations, consumers, employees, etc. as well as the resulting rebound effects (see Sect. 7.4).

Digitalization enables some types of work to be done from anywhere including home. This avoids commuting which can lead to significant CO_2 reductions. Global CO_2 emissions decreased by 17% in April 2020 as compared to 2019 during peak times—primarily due to less commuting but also due to lower production. Emissions have rebounded since then although a certain level of work from home still exists (WfH) [6]. The effect size of the advantage for climate change depends on the distance to work, the commuting mode, the type and amount of energy used for work at employee homes compared to employer sites and any associated behavior changes. The social impacts depend on the situation at home, the support for work-life-balance available, and the implications for social interactions. This includes enabling working from abroad, leading to development opportunities for remote areas. Increased efficiency of WfH has led companies to extend the scheme beyond pandemic shut-downs, however, this varies between company cultures, types of work to be done, etc. The resulting stress levels depend on the individual and their conditions. Digitalization of administration and work has impacts in terms of quantity and quality of jobs.

eCommerce positively impacts the access to products and services especially in remote or rural areas with the potential of lower costs for buyers. Overall, a higher consumption is linked to additional environmental stresses (on resources, emissions, waste, transportation, packaging, etc.) and social impacts in terms of quantity and quality of jobs, however, the effect sizes vary. Consumer platforms may also increase the number of shared products by digitally monitoring product use in real time (e.g. location, availability, use intensity).

Industry 4.0 or the fourth industrial revolution approaches are optimizing resource needs in smart factories in a technology-driven way. They utilize digitalization technologies such as the Internet of Things (IoT), Artificial Intelligence (AI), Machine Learning (ML), Machine Vision (MV), Data Analytics (DA) and many more. They use less resources, improve efficiencies, reduce waste streams, establish better emission controls, increase asset lifetimes by predictive and auto-planning maintenance activities, etc. [7] For example, IoT enables the real-time analysis of the condition, movement, and quantity of products and components. Digital interconnections between different value chain actors in carbon accounting and management may help reduce the product carbon footprint (e.g. by quantifying actual emissions through IoT and blockchain platforms, such as SiGreen/Estainium, Minepsider, Persefoni, CO_2NNEXTM or Catena-X). Going forward, industry 5.0—a value-driven approach—targets a sustainable, human-centric, and resilient industry. Industry 5.0 complements and extends the hallmark features of Industry 4.0 [8].

Digitalization can also help in the area of more sustainable cities. In a smart city information and communication technologies (smart systems) are used to improve citizens' quality of life and boost the effective use of resources.

Remanufacturing and durability improvements can be achieved by digitally tracking and tracing the availability and condition of functioning components in products. This may include circular service offers that collect user and product lifetime data digitally to provide product-life-extending services, take-back, pay-per-use, or pay-per-result services and/or to digitally optimize users' choices by guiding their product use (e.g. by incentivizing consumers of washing machines to reduce detergent, water, or energy consumption with a reduced service fee).

Digitalization enables industry to use less resources but also allows people to work or shop remotely which may have positive and negative environmental and social implications.

Optimistically, AI will help to generate wealth used to finance all people without requiring them to work. In the best case, this allows citizens to shift to more sustainable consumption and population growth patterns because nobody will need to impress with properties or require children to feed them in old age. Growth parameters may shift to dematerialized areas without energy demand—like education, social relationships, quality instead of quantity of products, using/sharing instead of owning, etc. AI may help in governance of all sustainability areas. There are of course also pessimistic scenarios regarding AI with respect to human rights, social sustainability, or related rebound effects compensating or outweighing their potential positive environmental and economic effects. [9–11]

A review of the impacts of AI on the 17 UN Sustainable Development Goals (SDGs) and their 169 sub-targets postulated that AI can enable 135 sub-targets but also inhibit 59 targets [12]. This means that AI can be both an enabler and blocker for some targets. In terms of social SDGs, AI may enable 67 (82% of social sub-targets) targets by supporting the provision of food, health (including research for new cures), water, and energy services through better/more reliable and faster data. On the other hand, AI may increase inequalities due to additional qualification requirements for any job or only serve the interest of the few companies, countries, or politicians it is being developed by. This may lead to negative impacts for 31 targets (38%). For economic SDGs, AI enables 42 targets (70%) mainly by increasing productivity and negatively impacts 20 targets (33%) by increasing inequalities and monopolies. AI supports 25 (93%) environmental SDG targets by better and faster data for climate modeling, biodiversity monitoring, identification of oil spills, and integrated energy systems (see Fig. 6.4).

On the other hand, AI and digitalization in general increase environmental impacts. This is mainly due to the significant energy consumption that leads to 3.5–4% of the world's carbon footprint. User activity contributes to this. There are estimates that one hour of video streaming leads to 90 g of CO_2 equivalent emissions if done via UMTS or G3, 13 g CO_2 in the case of G4 and 5 g CO_2 in the case of G5 while only 2 g CO_2 are produced if Wi-Fi is used [13]. Saving data in clouds, googling or using other internet tools or sending emails consumes energy. This demand may be ongoing if stored for years after being read. Programmers must be encouraged to write more energy efficient codes. This also demonstrates the influence of customer choices and the impact of infrastructure on the environment. Server rooms need a lot of electricity as they require constant cooling. The waste heat from server rooms can be used to heat offices, buildings, greenhouses, swimming halls etc. to reduce the negative impact. Overall, the total electricity demand of information and communications technologies could require up to 20% of the global electricity demand by 2030 an increase from around 1% today [14].

AI may support a majority of SDG targets by better & faster data for research, energy/food systems, and productivity. Risks for equality, human right, electricity consumption can be mitigated.

Similarly, autonomous vehicles (AV) have the potential to reduce vehicle emissions by optimizing routes or adjusting speed profiles to traffic lights. However, it is estimated that the required data connectivity and communication increases the emissions by 90 g CO_2 per km (to put this number into perspective: for newly registered passenger cars, the 2020 EU fleet target was 95 g CO_2 per km for all vehicle tailpipe CO_2 emissions) [15]. Again, renewable electricity can mitigate the impact.

The fundamental point is that digitalization has predominantly positive impacts on sustainability and there are sufficient options to mitigate the negative risks. These include using renewable electricity only and enacting more regulatory controls on the use of digitalization / AI to limit negative impacts on societies and human rights. Regulators also need to minimize the detrimental impact of improvements in technology, specifically within AI, on the market and in particular avoid the creation of monopolies, which disincentivize competition. For example, small and innovative start-ups cannot afford the latest technology (e.g. expensive chips) and hence do not have a chance in the market. This risk is documented in a yearly AI status report [16].

Similarly, platform work or the gig economy can carry risks of compromising workers' rights (e.g. fair pay, no excessive working hours), dependency on platforms and their algorithms, and exploitation of workers. However, they can also offer increased flexibility, independence, and additional income opportunities. EU is trying to mitigate the negative impacts by a regulation (Fig. 6.2).

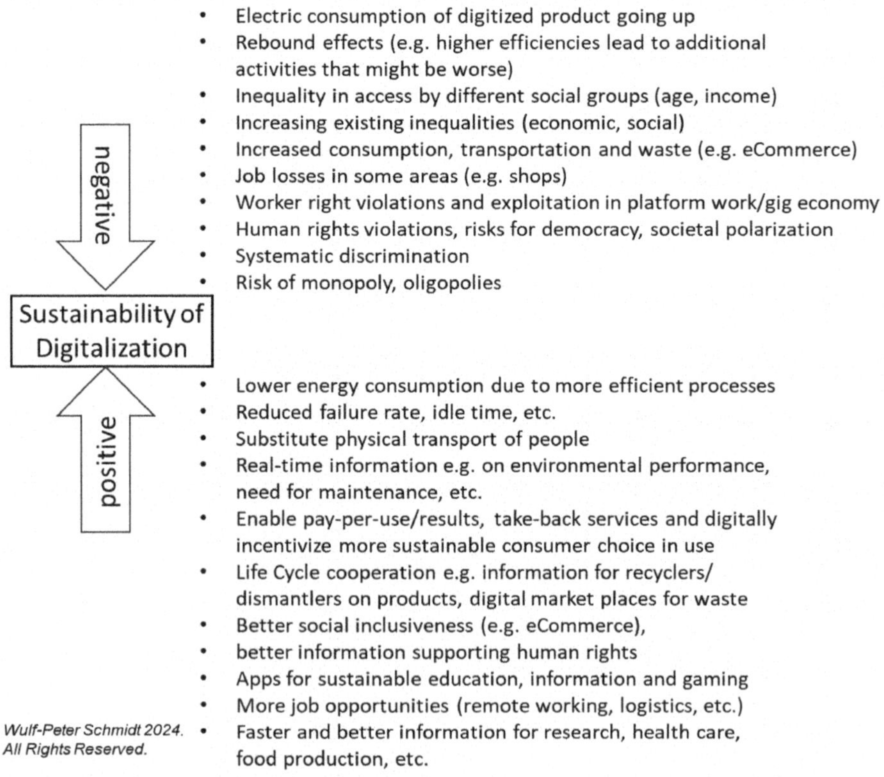

Fig. 6.2 Positive and negative impacts of digitalization on sustainability

Digitalization in itself is also a growing business when it comes to, for example, supporting the integration of sustainability processes and reporting within an organization. It is important, that the software packages offered are not seen as a solution in themselves, but that the organization must set up the right structures and processes first. A certain level of organizational maturity is a prerequisite to specific sustainability or ESG software adding value.

6.3 Sustainable Production Strategies—Example of EV Battery Production

EV battery production is an evolving technology due to the dynamic developments in battery chemistry and the limited state of maturity in production technologies. Figure 6.3 shows typical process steps required to establish EV battery packs. Basically, the anode (for Li-Ion batteries typically mainly graphite) and cathode (in case of Li-Ion batteries mainly Lithium or, depending on the battery chemistry, other elements) are created separately by mixing the anode/cathode active material together with solvents, binders (partly PFAS) and additives to form a slurry that is used to coat an aluminum or copper foil on both sides. This coated foil is dried in different temperature zones (50–160 °C). The dried foil is pressed and compressed by a rotating pair of rollers during calendaring and afterward slit into the final dimensions and stored in a vacuum oven to remove all remaining moisture and solvents. The electrode has now been manufactured and the cell assembly can start (separation of anode, cathode, and separator sheets from the roll material, stacking, packaging, electrolyte filling etc.). An important step is the formation of batteries where the cells are charged or discharged according to precisely defined current and voltage curves. This process triggers a strong evolution of gas and therefore must be degassed in a vacuum chamber. An aging process of up to three weeks ensures the final quality of the product. [17]

The current production of EV batteries is linked to high sustainability impacts in particular due to.

- High electricity demand. Roughly 50% of electricity is typically consumed in the formation step. Other high-consuming steps are air conditioning and controlled aging.
- High thermal energy demand (that is around 55% of all energy demand). Roughly 75% of natural gas is used in the mixing, coating, and drying steps.
- High water consumption in particular from mixing process and for cooling.
- High share of production waste due to inadequate batteries (quality issues).

The entire process is highly sensitive and the quality of the end product can easily be endangered by slight variations of process parameters (temperature, humidity,

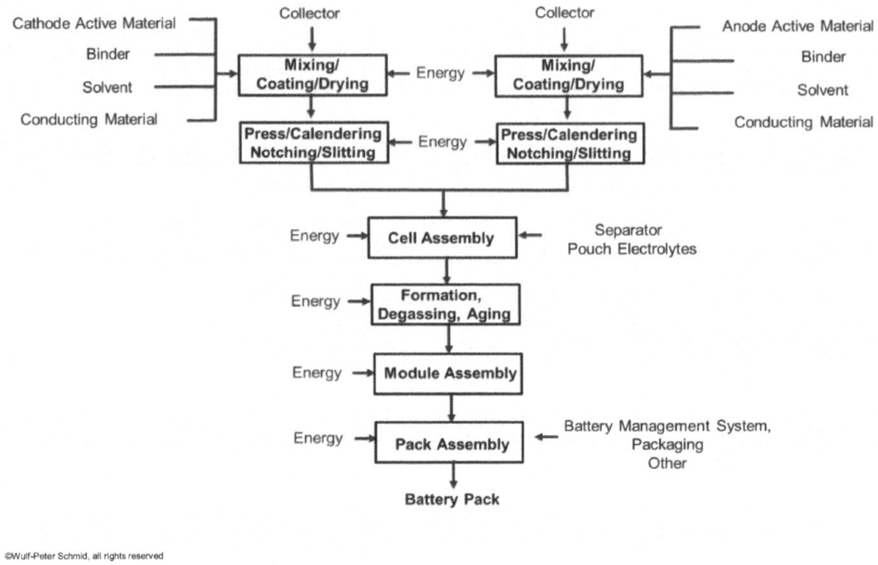

Fig. 6.3 Process steps in EV Battery production (based on [17, 18]). ©Wulf-Peter Schmid, all rights reserved

thickness, coating accuracy, porosity, edge accuracy, thermal/mechanical stress, electrostatic charging, small impurities, etc.). Thus, process improvements must be carefully tested and developed. However, given the impact of EV battery production, it is worthwhile to continuously improve the processes. There are different options for a more sustainable EV battery production including [17]:

- Optimize build-up phase:
  - Smaller dimensions of factories e.g. by electrification or by integrating processes like calendaring and slitting reduce the equipment footprint by 45%.
  - Greening buildings (facades and roofs with plants).
  - Local stakeholder involvement in planning workshops.
- Optimize input:
  - Different battery chemistry avoiding high-impact raw materials like Cobalt.
  - Replacing natural gas with non-fossil gas (see Sect. 6.5).
  - Electrification and use of renewable electricity by changing from thermal to electric energy (electrical heater, heat pumps).
  - Energy efficiency—for example, infrared and laser drying can reduce the process energy demand by 40%.
  - Dry coating (powder without solvents) leading to savings in later process steps as well.
- Improve efficiency:

  - Moving from batch to continuous processes—for example the slurry batch mixing can be transferred to a continuous extruder mixer, halving energy demands.
  - Improve precision e.g. by laser cutting—also reduces waste rate.
  - Process step integration—e.g. an integrated calendaring and slitting reduces process energy demand by 40%.
  - Avoid over-processing / outdated standards not needed anymore by deleting process steps—e.g. not drying cathode material (less energy, equipment footprint, less cost).
  - Reduce size—an example is to have an airtight structure around a machine and supply the dry air directly inside the machine instead of putting the machine in a dry air room.
  - Leaner production—reduce inventory stocks and make smart adjustments to processes with different process times creating waiting time and requiring buffers.

> EV battery production processes are sensitive but can be improved by various actions like continuous extruder mixing instead of batch processes, dry coating, infrared, or laser drying.

- Improve social effectiveness
  - Supply chain due diligence especially for raw materials like cobalt, nickel, and lithium.
  - Attractive wages, high working security standards.
  - Community involvement.
- Optimize output
  - Energy recovery and reuse of waste heat—for example energy recovery through powering HVAC system for high temperature aging via cell discharge or recovery heat from drying processes.
  - Water and solvent recycling—the cathode solvent NMP (N-Methyl-2-Pyrrolidone) can be recovered with up to 99% purity.
- Optimize build-down phase
  - Avoidance of toxic materials and fossil fuels in production reduces the risk of contamination.
  - Social plan for redundant employees and continued support for communities.

6.4 Low/Non-Polluting Production—Example Green Steel

To come as close as possible to non-polluting production, the chosen technologies must be optimized for efficiency, all energy used must be emission free and all products and waste streams must be part of a circular economy, ideally closed-loop recycling. Of course, these factors must be considered holistically, as the energy intensity required to close material loops and clear remaining emissions increases the fewer non-recycled and emission sources remain—a balance must be struck. Again, a holistic and extended life cycle perspective is important to avoid polluting parts being pushed outside of the system boundaries of the own production and therefore not being addressed.

Looking at greenhouse gas emissions, the biggest polluters in production are iron & steel as well as cement, mainly because these industries are not only energy intensive (due to high temperatures) but also because of direct CO_2 emissions in the material generation—due to the use of coal/coke as a reducing agent for iron ore or due to the release of CO_2 from Calcium Carbonate in the case of cement.

An example for the move towards low-polluting production is so-called "Green Steel". The term itself can be criticized as a misleading green claim (see Sect. 7.7.1), however, it is a good example for a transformational approach that moves from a high-impact material production to a lower impact material production that goes beyond incremental improvements or incremental recycled content changes. The production of steel and other metals is linked to around 10% of global primary energy demand [19]. The decarbonization glidepath for the steel sector suggests a linearly reducing carbon intensity from 1.74 t CO_2 equivalent per ton hot rolled steel to 1.2 by 2030 and 0.12 by 2050 [20]. The main approach for this significant reduction in CO_2 emissions is the replacement of coke as the reduction agent by green hydrogen as well as using electric furnaces to allow an increasing use of steel scrap. Classification systems are evolving to suggest that maximum CO_2 emissions allowed per ton of steel should depend on their recycled content [21]. The main limitations are the availability of green hydrogen and scrap iron and steel because all available iron and steel scrap is usually already recycled for economic reasons. Also, establishing the production and infrastructure capacities for sufficient green hydrogen takes time.

Replacing coke as the reduction agent & energy carrier is key for low CO_2 iron & steel. Green hydrogen or bio-oil are suitable alternatives that require sufficient infrastructure.

For iron—the precursor of steel—other ideas exist, including some that claim to enable a net negative carbon iron generation [22]. For example, Charm Industry uses "bio-oil" for iron creation. "Bio-oil" is made from organic waste like corn stover, rice straw, and timber slash that is heated up for a few second for pyrolysis (500 °C). The assumption is that 1 ton of biomass entered the "bio-oil" production removes around 0.85 net tons of CO_2 from the atmosphere (1 ton biomass contains 1.65 tons

Table 6.2 Lower carbon emission alternatives for reducing iron ore for iron production

Reducing agent for iron ore	Resulting carbon footprint	Challenges
Coal (conventional)	Base value	Expensive CCS* needed to get to net zero
Natural gas	Reduced	Still fossil based, CCS* needed
Hydrogen	Carbon neutral possible**	Green hydrogen availability, cost
Bio-oil	Carbon neutral or maybe even carbon positive possible	Scale-up, transport distance of bio-oil
Electrolytic	Carbon neutral possible**	Theoretical, no carbon in iron

* CCS—Carbon Capture and Storage; ** if based on renewable electricity only
©Wulf-Peter Schmid, all rights reserved

CO_2, fast pyrolysis emits 0.78 tons CO_2 alongside other emissions of production and transport amounting to 0.021 tons CO_2)—note: this calculation is not based on a system extension approach as introduced in Fig. 2.5 and Fig. 2.7. Bio-oil is a mixture of "water, organic acids, aldehydes, ketones, sugars, phenols, and many other organic compounds derived from the thermal breakdown of cellulose" but cannot sustain a flame due to its lower energy density than crude oil. Therefore, it is less useful for energy generation but its carbon content is of potential interest for iron making (or carbon removal). This "bio-oil" is gasified (to hydrogen and carbon monoxide) to enable it to replace coke in iron ore reduction. The resulting emissions are—compared to traditional coke-iron ore process—more CO_2 heavy but without/with less impurities like sulfur that would make subsequent carbon sequestration less efficient. Around 3 tons of "bio-oil" are needed per ton of iron and 1.2 tons of CO_2 would need to be sequestered again at the end of the process. The scale-up would take until 2030. Due to the transport-sensitivity of "bio-oil" (low energy content makes transportation less efficient), a mobile production of bio-oil is in the works. [22] (Table 6.2).

6.5 From Reducing & Reusing Energy to Energy System Integration

ISO 50001:2018 requires organizations to determine, analyze, and optimize energy consumption systematically and continuously. In this context, energy efficiency is an important performance indicator. In fact, the International Energy Agency and the International Renewable Energy Agency suggest a doubling of the global average annual rate of energy efficiency improvements from around 2% to over 4% every year until 2030 to limit warming to 1.5 °C [23]. Energy efficiency is the ratio between the resulting output or yield of services, goods, commodities, or energy and the energy used (input). The higher the efficiency, the lower the energy loss is.

It is estimated that 29% of the total industrial energy demand is currently lost as waste heat [24]. The amount of (renewable) energy needed can be reduced significantly if waste heat recovery is done by one of the following methods:

- Direct heat provision by heat exchangers. There are various examples of how the waste heat of one process can be used for another process on a slightly lower temperature level directly. This can be done especially in so-called integrated industries that produce and demand heat / different temperature levels like food and beverage, paper, steel, ceramic, and many more industries. Alternatively, the waste heat can be used as district heating to support neighboring communities. This energy flow management is easier to organize within one organization or at least within the same industrial site compared to energy flows shared between different organizations and locations.
- Indirect heat provision by storage systems. The heat can be stored and used at a later point in time or transported to other consumers. Energy storage technologies include batteries, pumped hydro, liquid air energy storage (LAES), molten salt energy storage (MOSAS), or conversion into hydrogen.
- Indirect thermal energy provision by heat pumps can increase efficiency as the "coefficient of performance" (i.e. the ratio of energy delivered by the heat pump in terms of heat output from the condenser related to the power supplied to the heat pump compressor) is commonly significantly above 1.
- Energy recovery via transformation into electric power. Thermal waste energy can be converted to electricity. An example for the conversion to electricity is the Organic Rankine Cycle (ORC). The cycle starts with a heat exchanger where a working fluid evaporates under the effect of heat and develops steam pressure. The pressure is released via a gearless micro turbine, which drives an integrated generator that produces the electric current. After expansion, the steam enters a second heat exchanger, the condenser, where the steam is cooled using the ambient air so that the steam liquefies and is fed back into the cycle as a working medium. There are around 100 TWh of unused waste heat between 100 and 200 °C in Europe—e.g. waste heat in food industry, large-scale laundry, paper machines, plastic processing, emission control systems, and ovens. If all these sources would be used to generate electricity, around 7 TWh electricity could be generated—without any additional direct emissions. The potential is even higher if higher-temperature waste heat is also used [25].
- Energy recovery via steam. Combined Heat and Power (CHP) plants convert heat into steam to generate electricity like normal energy plants. In addition, the steam is used to generate (district) heat.

> Thermal energy can be recovered directly by heat exchangers, indirectly by storage or heat pumps, or used for electricity generation (Organic Rankine Cycle, Combined Heat&Power).

Of course, it is better to reduce the energy consumption in the first place by measures like:

- Smarter consumption:
  - Turning on fans, ventilation systems, pumps, dryers, etc. only when and for as long as needed.
  - Heat curve optimization—i.e. instead of cycles with heating followed by cooling and re-heating again, smoothing the temperature curves. Examples include sintering heating curves where the aging process is replaced by inspections after high temperatures subside
- More energy efficient technologies and processes (e.g. LED lighting, more efficient pumps etc.)
- Integrating, replacing, deleting energy intensive processes

By widening the perspective beyond the individual production, a broader energy system integration is possible. This entails the planning and operation of a holistic energy system across multiple energy sources, infrastructures, and consumer sectors with the aim to establish low-carbon, reliable, and resource-efficient energy sources through creating stronger links. The priority is always the least energy intensive option, unavoidable energy waste streams are reused and synergies are exploited across all sectors [24]. The main characteristics of sustainable energy system integration are [26] (Fig. 6.4):

- A more efficient and "circular" system, where waste energy is captured and reused.
- A cleaner power system, with more direct electrification of end-use sectors such as industry, heating of buildings, and transport.
- A cleaner fuel system for hard-to-electrify sectors like heavy industry or heavy goods transport.

Currently, systems are not only linear but also centralized and not sufficiently interconnected. Future integrated energy systems are more resilient by having.

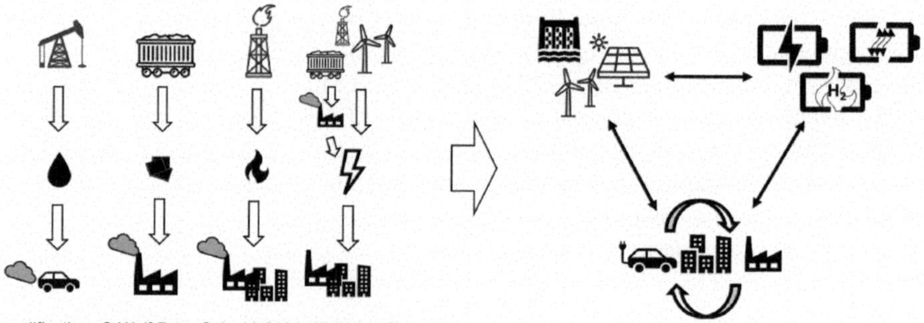

modifications © Wulf-Peter Schmidt 2024. All Rights Reserved.

Fig. 6.4 From traditional linear energy flows to an integrated energy system ([24, 26] and own modifications). Modifications © Wulf-Peter Schmidt 2024. All Rights Reserved

- Decentralized renewable energy sources
- Storage systems (power, power-to-gas, heat); stationary and mobile
- Smarter and "productive consumers" that not only adjust their demand to balance peak-demand times but also generate energy themselves (heat from industry, net positive buildings, bidirectional charging of vehicles, etc.).

Decentralization can also help accelerate the growth of renewable energy infrastructure by complementing large offshore and onshore wind projects or huge solar farms with decentralized projects that are embedded into existing infrastructure and can be implemented quicker. These include start-ups that use mobile phone masts and rooftops to add small wind turbines [27] or mobile wind power technologies that use high(er) altitudes without relying on fixed installations that require long approval times [28].

A sustainably integrated energy system is more resilient by having strong links between decentralized renewable sources, stationary/mobile storage & smarter, productive consumers.

6.6 Renewable Energy in Production

Roughly three quarters [29] to 86% [30] of human greenhouse gas emissions are related to energy. A strategic switch to renewable energy not only addresses the majority of these emissions but is also economically positive given the significantly lower lifetime costs of new solar and wind energy infrastructure that are linked to only a sixth of fossil fuel costs—at least looking at costs in Europe during the 2022 energy crisis [31]. "Onshore wind is expected to have, on average, the lowest levelized costs of electricity generation in 2025" when comparing all energy sources and has the lowest cost in at least 10 of 14 analyzed countries [32]. In 2022, the global net electricity growth was already primarily met by renewable sources [30]. The target agreed at COP 28 is to triple the world's installed renewable energy generation capacity to at least 11,000 GW by 2030 [23] (Fig. 6.5).

The levelized costs of generating electricity (LCOE) represent the discounted constant lifetime remuneration of the supplier of electricity during its lifetime taking into consideration capital, maintenance, and operational costs including fuel and carbon costs as well as decommissioning costs at end-of-life. However, LCOE "indicates the costs of a generic technology, not the specific costs of a specific project in a specific market" [32]. Where a generation of renewable electricity on-site is not possible the next best option is long-term Power Purchase Agreements (PPAs) that provide third-party investors with stability for their renewable electricity projects as well as supply certainty for the buyers.

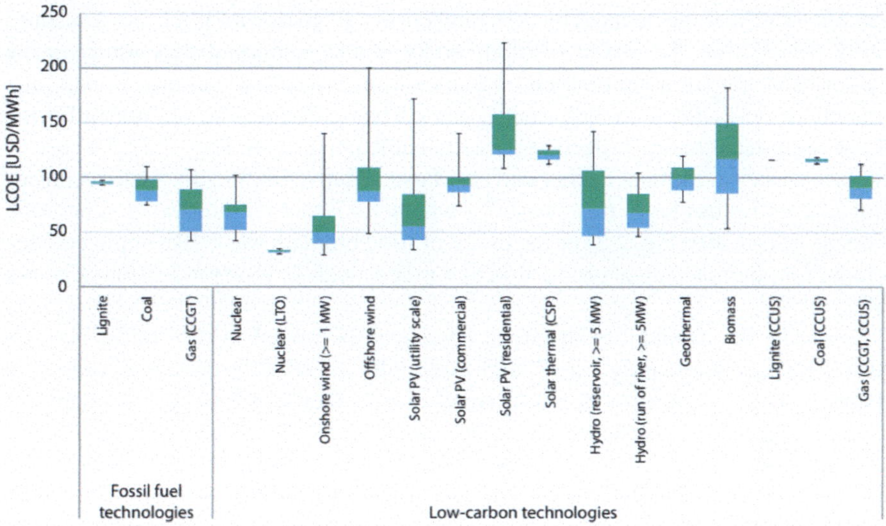

Fig. 6.5 Levelized costs of generating electricity (LCOE) per technology—Values at 7% discount rate. Box plots indicate maximum, median, and minimum values. The boxes indicate the central 50% of values, i.e. the second and the third quartile [32]

> Onshore wind power has the lowest levelized costs in most analyzed countries. Long-term Power Purchase Agreements are the best option if on-site generation is not possible.

On-site generation of renewable electricity requires storage systems—these can include stationary batteries (including second-life Eletric Vehicle batteries) or Green Hydrogen.

The switch from thermal energy to renewable electricity is a generic method of decarbonizing production. For example, natural gas-powered thermal heaters can be switched to electrical heaters that are linked directly or indirectly via hot oil boilers to heat exchangers of up to 350 °C or more. Renewable electric solutions are also possible for higher temperatures, for example for electric smelting (1300°) and sintering (1050 °C) in the magnet production.

Where a switch from thermal energy to electricity is not possible or reasonable due to a lower efficiency at high temperatures/fast temperature changes, other gasses like biogas or Green Hydrogen are an option. However, switching thermal energy to renewable sources is difficult in the present due to limited availability—the International Energy Agency (IEA) estimates that the current biogas potential covers only 20% of the demand for natural gas and meets 10% of the demand with the least expensive biomethane option. This option is twice as expensive in Europe, 4 times in the US and 1.2 times in developed Asia. In the future, however, availability will increase to 50% of the demand by 2040 and costs will go down [33]. From a sustainability perspective, the aspect of biodiversity is also important depending on the sources of biogas (or other biofuels). Monocultural agriculture techniques with high amounts of pesticides may help reduce net carbon emissions but do not fulfill the "do no significant harm" criteria for a sustainable choice (see Sect. 3.5). The use of biomass waste for thermal energy can also have negative implications when considering alternative usages of this "waste"—whether it is carbon removal or generation of compost (see also Fig. 2.7) or the usage of food biomass waste like apple pomace as input for pharmaceutical production of nutraceuticals or food production of pectin [34].

Similarly, the availability of Green Hydrogen is limited in the 2020s. Today's hydrogen is predominantly "gray" and essentially generates no real environmental advantage (see Fig. 6.6 for different types of hydrogen). In Europe, the supply of "green" hydrogen is estimated to be less than 1% of final energy in the EU by 2030 and globally by 2035. There is a potential for a share of 3.2% to 11.2% in EU and 0.7% to 3.3% globally by 2040 [35] and 18% in EU by 2050 [36]. Green Hydrogen will be 5 times more expensive than natural gas in Europe by 2025 (blue hydrogen 3 to 4 times more expensive depending on how much carbon is captured). The natural gas-based blue, gray, or turquoise hydrogen forms do not address the dependency on suppliers of natural gas and have the risk of methane leakage (high greenhouse gas impact). Green hydrogen will be the cheapest form of hydrogen by 2035–2040 [37] and is seen as an important method of energy storage from intermittent renewable electricity sources—within the regional grid but also for own/on-site renewable electricity generation.

> Where electrification is no option, green hydrogen or biogas are alternatives to decarbonize thermal energy. Constraints include availability, biodiversity & lost biomass for biogas.

The often-debated nuclear electricity as a carbon-free alternative is—irrespective of debates around inherent safety risks from terror attacks or other risks and the still unanswered disposal questions—not a viable option to cover incremental energy needs given the long planning and development times as well as the significant higher costs compared to renewable electricity even if renewables are coupled with storage systems.

6.7 Cement and Sustainable Building

The manufacturing of Portland cement to be used as a binder in concrete is a major contributor to greenhouse gas emissions and accounts for about 4–7% of worldwide total human emissions. This high impact has two reasons: Firstly, the use of limestone

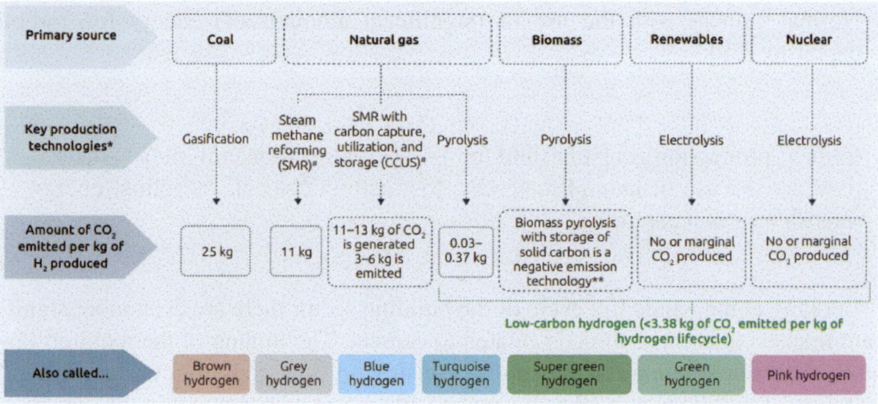

Fig. 6.6 Different pathways of hydrogen production vary in carbon intensity (reprint from Capgemini Research Institute [36]). *Note* * The pyrolysis of biomethane, biomass, waste, or wastewater, with subsequent storage of solid carbon, is a negative-emission technology, since the CO_2 previously removed from the atmosphere and neutralized in the biomethane is not released during the pyrolysis reaction or use of the hydrogen produced and, consequently, no climate-damaging greenhouse gas effects are produced. ** CO_2-eq emissions could be higher than indicated above when considering the efficiency of the CCUS process, methane leakage in the system, and the time horizon between 20 and 100 years for the Global Warming Potential (GWP)

means that encapsulated CO_2 is released. Secondly, the high temperature of around 1500 °C of the rotary kilns leads to very high energy demand. Cement, together with water, sand, and gravel, is used for concrete. The increasing demand for concrete from the construction and building industry has tripled the extraction of non-metallic minerals, especially sand and gravel. In consequence, sand is the world's most used resource after clean water. Sand and gravel extraction disrupts water supplies, hydrological functions as well as river and coastal ecosystems [19]. There are different alternatives for the current traditional cement/concrete including:

- Non-cement / non-concrete: natural material like bamboo, wood. On one hand, the use of natural materials has the advantage that their use in buildings stores the incorporated carbon over decades. However, the sustainability of using natural materials mainly depends on how sustainable their sourcing is, considering biodiversity implications and the condition of forests in most countries as well as their alternative usages.
- Low-carbon cement is not based on CO_2-containing limestone but on a "wide variety of silicate rocks and minerals as raw material feedstock… including clay waste streams". This feedstock is broken up in mills and vitrified into a glassy powder suitable for blending in cement products (up to 40%). The process may lead to up to 70% less CO_2 and 90% lower NO_x emissions than Portland Cement. [38]
- Recycling material—cement is one of the biggest sources of waste streams in several countries.

- Normal cement with the use of (significant amounts of) energy for carbon capturing [39].

Cement production is responsible for 4–7% of all carbon emissions. Alternatives are the use of natural materials, low-carbon cement, recycling, energy intensive carbon capturing.

Looking at the whole life cycle of the building sector there are even more significant impacts than those directly related to cement. The mining of the required raw materials, the production and transport of construction materials, the construction and operation of buildings as well as the end-of-life waste management in the construction and demolition phases cause around 40% of global greenhouse gas emissions [19]. The most important aspects in the assessment of the sustainability of buildings are shown in Fig. 6.7.

- Sustainable energy sources (e.g. solar cells).
- Sustainable sources building material and production (see above)
- The conservation and efficiency of energy, resource, land, and water use.
- The safety of building materials including protecting occupant health.

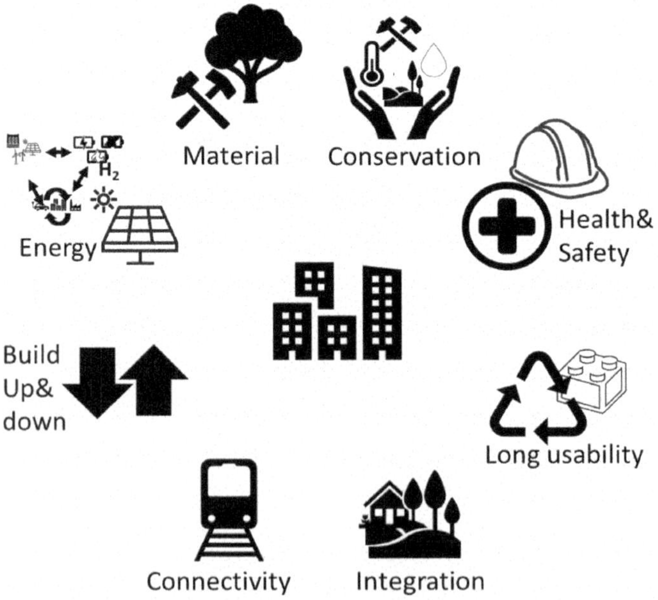

Fig. 6.7 Sustainable buildings. ©Wulf-Peter Schmidt, all rights reserved

- The durability, modularity, reusability, ease of repair, upgrading, repurposing and refurbishment of materials and buildings, and use of recycled materials.
- The architecture and site of a building with consideration of its impact on land use, the environment, employee productivity, mobility needs and implications, neighborhood, and the society (social aspects and livable communities).
- The architecture and site of a building with consideration of its functioning in the surrounding infrastructure context (mobility, connectivity, energy system integration).
- The impacts during built-up and down.

The first point (energy source) includes not only typical aspects like solar panels, heat pumps, and a system energy integration (Sect. 6.5) but also maximizes the use of natural daylight, ventilation, thermal, and acoustic insulation. Building materials have been addressed above and sustainability considerations are mainly driven by substituting classical concrete.

The third point (conservation) is relevant from an energy perspective because most buildings are use-phase sensitive (see Sect. 2.1.2) due to their commonly long lifetime and the high energy demand for cooling and heating (exemptions like parking garages etc. excluded). In consequence, thermal insulation properties may outweigh higher impacts during the production phase. Nevertheless, refurbishment, renovation, upgrading, etc. are often more sustainable than constructing a completely new building from scratch—as long as the upgrade can establish a sufficiently high standard in terms of thermal insulation and heating/cooling systems—see also the fifth item. The conservation of water refers to both reducing water consumption but also absorbing excess water from outside. The reduction of the building dimensions and footprint can help conserve resources and minimize land use.

The fourth item (health & safety), for example, is important when it comes to the use of substances in insulation materials as well as the impact of insulation on interior air quality. Safety of buildings should be a given but events like the collapse of the Rana Plaza textile manufacturing site in Bangladesh in 2013 underline their importance. Alternative materials like compressed earth bricks—made from soil, natural fibers, and clay—can last for centuries, are easy to repair and have a very low-carbon footprint (95% less than traditional concrete block constructions [19]).

The fifth item enables the implementation of circular strategies in the building sector like the R10 principles (compare Table 4.2) and ensures that buildings can be used for longer, keeping the invested resources, energy, and capital useful for longer.

Greening of buildings can limit the impact of land use and improve biodiversity as well as insulation of the building. In general, the integration of the building into the surrounding landscape is key for acceptance and ensures livable communities.

The site of buildings is also relevant from a different perspective: if the surrounding infrastructure is ignored, the functionality of the building will suffer from sub-optimal mobility, energy/water supply as well as digital connectivity.

Sustainability of buildings depends on the energy & material sources and their conservation, health & safety for occupants/workers, circularity, site as well as the built-up & down phases.

Last but not least, the construction phase as well as the deconstruction phase must be planned carefully to avoid unnecessary impacts on the environment and communities.

Sustainable cities extend far beyond sustainable buildings. The Sustainable Development Goal (SDG) 11 targets an "intelligent urban planning that creates safe, affordable and resilient cities with green and culturally inspiring living conditions" [40]. Resilience is required to address environmental, social, and economic challenges as well as enable smart mobility planning.

Even without climate change, cities are vulnerable to high temperatures. City planning must ensure that sufficient ventilation is achieved by cooler, green areas (parks, meadows, forests) that extend from rural regions straight into the city, ideally crossing through the center of city to the other end of the city. This can be accompanied by an increase in the albedo of surfaces like roofs (unless these are already being used for photovoltaic energy generation), greening of facades or roofs as well as water installations (rivers, fountains, etc.). Resilience against flooding from heavy rain requires a migration of water from sealed areas (roads, buildings, etc.) to permeable areas that can take up rain like huge sponges. Surplus rainwater must be directed to water buffer areas. The urgency of these measures has been stressed by various organizations and individuals including the Intergovernmental Panel on Climate Change (IPCC) [41]. The IPCC stated with high confidence that extremely hot periods have intensified in cities already with the adverse impacts concentrated among economically and socially marginalized urban residents. In 2023, Paris simulated a heat wave temperature peak of 50 °C in 2032.

City planning must also foster social equality but avoid segregation of different income groups. The SDG 11.3 sub-target demands inclusive, participatory, integrated and sustainable human settlement planning and management. Inclusion can be fostered through public hearings, community workshops or participatory decision-making tools. Affordable housing can be enabled by financing, cooperative and social housing—preferably creating areas containing all income groups. Public investments in sanitation, public spaces, parks, cultural and exercise areas as well as mobility will support a more sustainable city.

In the mobility space, cities are experimenting with low or even zero-emission zones where high-emitting vehicles are banned or financially disadvantaged. Car-free zones have been implemented as well. A significant reduction of parking spaces impacts mobility preferences most effectively and can be accompanied by lower speeds and safe bicycle lanes. A better mix of living, working, and shopping areas with small distances in between them and good public transport links as well as more parking opportunities outside cities accompanied by multi-modal mobility concepts are longer term measures.

Digitalization can also help create more sustainable cities. In a smart city, information and communication technologies (smart systems) can be used to improve citizens' quality of life and boost the effective use of resources.

Sustainable cities are resilient to climate change and heavy rains due to green lungs, have mixed-income working & living areas, multi-modal mobility, and smart information systems.

A ranking of sustainability of 70 global cities [42] considers various criteria. The biggest differences between cities have been seen for the criteria emissions per capita, vehicle dependency, sustainable transport, and open space. This ranking therefore suggests improvement potentials especially in these areas, however, not all important indicators—in particular in the social area—are covered in the ranking. Interestingly, the top 4 most sustainable cities are all Scandinavian. Similar results are seen in other rankings [43].

6.8 Sustainable Food Production/Agriculture

Food production is the single largest driver of the ecological impact of humans and the transgression of planetary boundaries because agriculture covers two fifths of the planets usable land surface and therefore massively impacts biodiversity. In addition, the sector is responsible for one-third of all Greenhouse gas emissions (8–10% of all Greenhouse gas emissions is the result of lost and wasted food), 70% of global freshwater withdrawals and one-third of global nitrogen mobilization [19]. In several countries, worker's rights and health and safety are compromised. Structural risks of child labor and other social issues further limit the sustainability of this sector.

However, sustainable food production includes various options such as:

- Transition from meat and dairy products to plant and insect-based food products –commonly reducing the energy consumption and emissions by a factor of ten due to the consumption and growth of animals. Insect foods like cricket protein more efficiently transfer input into the resulting valuable nutrients and use one-sixth less feed (low-value agricultural waste), 1/1,000th the amount of water, and 1/20,000th the amount of land compared to the same amount of beef. [19]
- More sustainable harvesting cycles and regenerative agriculture like less overfishing, more crop rotation like three-field rotation farming with a closer nutrient loop, or farming approaches where different species grow in the same area complementing and supporting each other (polycultures, permacultures, biodynamic farming, agroforestry).
- No or reduced irrigation, artificial fertilization, and pesticides.

- Food and harvest waste avoidance and/or circular use of unavoidable waste. This includes "management of transport and storage, more refrigeration and smart planning, and technology at the consumer and food service levels" [19] but also packaging as well as consumer behavior changes.
- Precision agriculture can reduce the need for pesticides by up to 99% and is based on advanced digitalization, artificial intelligence, and automated robotics [19].
- Use of renewable energy sources.
- Integrated pest management that uses natural predators to control populations of pests.
- Hydroponics and Aquaponics—In hydroponics, root systems are placed directly in a defined mineral solution or in gravel and the needed nutrients are added to the water given to the plant, removing the need for soil and optimizing the amounts of nutrients.

  Aquaponics combines raising aquatic animals with the growing of hydroponic crops. In this approach, the waste water from the animal is used to nourish the plants. Since the water contains nutrients from the animal waste, it can be used to feed the plants. It is then recycled back into the system to be used once again by the fish [44].
- Vertical or urban farming using less land and optimizing water consumption can improve the sustainability footprint as long as only renewable energy sources are used [45]. Otherwise, studies have shown that urban farming has a six times higher carbon footprint than conventional farming [46].
- Low water farming based on drip irrigation and scheduling irrigation to reduce evaporation losses, capturing and storing water, planting plants resistant to droughts as well as covering soil with compost and mulch, etc.
- Better human and worker's rights auditing but also fairer contracts between farmers and the food industry are key elements in improving the social implications.

Agricultural issues around biodiversity, carbon, water, land, or social footprint can be addressed by various farming strategies, substituting meat, better management and governance.

6.9 Carbon Removal

For processes that cannot be transformed to avoid any pollution, there is the option to capture, remove, and store the emissions. At least for CO_2, this is a widely discussed opportunity—although not widely used yet (ISO 14068:2023).

"Anthropogenic activities removing Carbon Dioxide (CO_2) from the atmosphere and durably storing it in geological, terrestrial or ocean reservoir' are defined by the International Panel of Climate Change (IPCC) as Carbon Dioxide Removal (CDR). It includes existing and potential anthropogenic enhancement of biological or geochemical sinks and direct air capture and storage, but excludes natural CO_2 uptake not directly caused by human activities." [47]

It is important that the carbon removals are real and potential significant adverse impacts to the environment and society are minimized [48]. The durability of different carbon removal pathways differs:

- Direct air capture and sequestration uses chemical processes to divert CO_2 directly from emissions or ambient air based on large fan-like machines injecting the CO_2 into geological reservoirs (typically for 1000 + years). This option is discussed especially for big emission streams with high CO_2 content (e.g. from cementation of calcium carbonate).
- Enhanced weathering & mineralization accelerates, for example, the natural process of rock weathering or stores the captured CO_2 in stable mineral forms for 1000 + years.
- Ocean CDR involves capturing and storing CO_2 directly from the atmosphere into the deep ocean e.g. by adding alkaline substances that enhance ocean alkalinity leading to the ocean's increased ability to absorb and store CO_2 for 1000 + years.
- Biomass CDR uses biomass or seaweed to remove CO_2 from the atmosphere via photosynthesis and store it through combustion (BECCS), gasification (bio-oil), pyrolysis (biochar) or by letting the carbon sink in deep water. The durability of biomass CDR is usually between 300 and 1000 years.

All these methods are usually energy intensive (especially direct air capture and enhanced weathering), expensive and the scaling-up is still under development. The IPCC states that the "large-scale deployment of carbon dioxide removal (CDR) … (is) uncertain and entail(s) clear risk" and stresses that "CDR deployed at scale is unproven, and reliance on such technology is a major risk in the ability to limit warming to 1.5 °C" [47]. On the other hand, the IPCC identified that almost all pathways left to limit climate change to 1.5 °C above preindustrial levels assume the use of CDR—however, their usage is limited to the completely unavoidable remaining emissions where no better alternatives exist.

> Carbon removal methods include direct capture & sequestration, enhanced weathering & mineralization, deep ocean storage, and biomass. Main risk is the unproven scaling capability.

References

1. DIN, 8580-2022:12, Manufacturing Processes—Terms and Definitions, Division
2. A. Leidena, P.-J. Brand, F. Cerdas, S. Thide, C. Hermann, Transferring Life Cycle Engineering to Surface Engineering. Procedia CIRP **90**, 557–562 (2020)
3. Lowell Center for Sustainable Production, Sustainable Production Defined. https://www.uml.edu/research/lowell-center/about/sustainable-production-defined.aspx. Accessed 28 Mar 2024
4. Renault, ReFactory (Renault, 25 Nov 2020). https://www.renaultgroup.com/en/news-on-air/news/re-factory-the-flins-site-enters-the-circle-of-the-circular-economy/. Accessed 8 Apr 2023
5. A. Neligan, C. Schleicher, B. Engels, T. Kroke, Digitaler Produktpass—Enabler der Circular Economy. Relevanz und Umsetzbarkeit durch Unternehmen, Institut der deutschen Wirtschaft Köln e. V. (Berlin/Köln, 2023)
6. G. Shreedhar, K. Laffan, L.M. Giurge, *Is Remote Work Actually Better for the Environment?* (2022)
7. M. Javaid, A. Haleem, R.P. Singh, S. Khan, R. Suman, Sustainability 4.0 and Its Applications in the Field of Manufacturing. Internet Things Cyber Phys. Syst. **2**, 82–90 (2022)
8. X. Xu, Y. Lu, B. Vogel-Heuser, L. Wang, Industry 4.0 and Industry 5.0—Inception, Conception and Perception. J. Manuf. Syst. **61**, 530–535 (2021)
9. MIT Technology Review Insights, The Great Acceleration: CIO Perspectives on Generative AI (2023)
10. S. Falk, A. v. Wynsberghe, Challenging AI for Sustainability: What Ought It Mean? *AI Ethics* (2023)
11. A.K. Kar, S.K. Choudhary, V.K. Singh, How Can Artificial Intelligence Impact Sustainability: A Systematic Literature Review. *J. Clean. Prod.* **376** (2022)
12. R. Vinuesa, H. Azizpour, I. Leite, M. Balaam, V. Dignum, S. Domisch, A. Felländer, S.D. Langhans, M. Tegmark, F.F. Nerini, The Role of Artificial Intelligence in Achieving the Sustainable Development Goals. *Nat. Commun.* (2020)
13. F. Schätzing, Was, Wenn Wir Einfach Die Welt Retten? Handeln in Der Klimakrise (Köln, Kiepenheuer & Witsch, 2022)
14. N. Jones, How to Stop Data Centres from Gobbling Up the world's Electricity. Nature **561**, 163–166 (2018)
15. R. van Oosterhout, P. Striekwold, M. Wand, On Data-Induced CO_2 Emissions of Vehicle Automation: An Overlooked Emission Source. Sustainable Horizons **9**, 100082 (2024)
16. N. Benaich, *State of AI Report* (Air Street Capital, 2023)
17. H. Heimes, A. Kampker, C.L.M. Lienemann, *Lithium-Ion Battery Cell Production Process* (VDMA Battery Production, 2019)
18. H.C. Kim, S. Lee, T.J. Wallington, Cradle-to-Gate and Use-Phase Carbon Footprint of a Commercial Plug-in Hybrid Electric Vehicle Lithium-Ion Battery. *Environ. Sci. Technol.*, 11834–11842 (2023)
19. Circle Economy, *The Circularity Gap Report 2023* (Circle Economy, Amsterdam, 2023)
20. Global Steel Climate Council (GSCC), The Steel Climate Standard—Framework for Steel Product Certification and Corporate Science-Based Emissions Target (2023)
21. R. Geres, D. Holzner, S. Weigert, M.B.W. Beckmann, R. Winter, Vorschlag für ein Regelbuch Klassifizierungssystem Grüner Stahl—Version vom 31. März 2023 (Witrschaftsvereinigung Stahl, Berlin, München, Langenbach, 2023)
22. Charm Industrial; Plus Explanation of Sebastian Smith of Charm Industrial. https://charmindustrial.com/faq.
23. COP 28 UAE, Global Renewables and Energy Efficiency Pledge. https://www.cop28.com/en/global-renewables-and-energy-efficiency-pledge. Accessed 17 Dec 2023
24. U. Vögtle, *Integrated Energy Production—White Paper* (MAN Energy Solutions, SE, Augsburg)
25. SOOMIQ, PowerQube und ORC-Technologie, SOOMIQ. https://www.powerqube.net/powerqube-und-orc-technologie.

26. European Commission, *EU Energy System Integration System* (Brussels, 2020)
27. MOWEA, Modular Wind Energy Systems. https://www.mowea.world/en/.
28. Enerkite, Green Energy from High Altitudes. https://enerkite.de/en/
29. H. Ritchie, M. Roser, P. Rosado, CO_2 and Greenhouse Gas Emissions. *Our World in Data* (Aug 2020)
30. UNEP, *Broken Record—Temperatures Hit New Highs, Yet World Fails to Cut Emissions* (again) (United Nations Environment Programme, Nairobi, 2023)
31. International Renewable Energy Agency, *Renewable Power Generation Costs in 2021* (IRENA, Abu Dhabi, 2022)
32. International Energy Agency, *Nuclear Energy Agency, Projected Costs of Generating Electricity* (IEA, Paris, 2020)
33. International Energy Agency, *Outlook for Biogas and Biomethane—Prospects for Organic Growth* (IEA, Paris, 2020)
34. N.R. Putra, D.N. Rizkiyah, A.H.A. Aziz, M.A.C. Yunus, I. Veza, I. Harny, A. Tirta, Waste to Wealth of Apple Pomace Valorization by Past and Current Extraction Processes: A Review. *Sustainability* **15**(830) (2023)
35. A. Odenweller, F. Ueckerdt, G.F. Nemet, M. Jensterle, G. Luderer, Probabilistic Feasibility Space of Scaling up Green Hydrogen Supply. Nat. Energy **7**, 854–865 (2022)
36. Capgemini, *Low Carbon Hydrogen—A Path to a Greener Future* (Capgemini, 2023)
37. F. Ueckerdt, P.C. Verpoort, R. Anantharaman, C. Bauer, F. Beck, T. Longden, S. Roussanaly, On the Cost Competitiveness of Blue and Green Hydrogen. *Preprint Nature Portfolio* (Mar 2022)
38. Terra CO_2, Decarbonizing Concrete by Replacing Cement. https://terraco2.com/.
39. Heidelberg Materials, Heidelberg Cement to Build the World's First Carbon-Neutral Cement Plant (2021). https://www.heidelbergmaterials.com/en/pr-02-06-2021.
40. United Nations, SDG, 25 Sept 2015. https://documents-dds-ny.un.org/doc/UNDOC/GEN/N15/291/89/PDF/N1529189.pdf?OpenElement.
41. IPCC, Synthesis Report of the IPCC Sixth Assessment Report (AR6) Summary for Policymakers (Intergovernmental Panel on Climate Change Geneva, 2023)
42. R. Torrie, N. Morson, Corporate knights: Sustainable Cities Index Report (2023)
43. CEMEX Ventures, These are the Most Sustainable Cities in the world (2023) https://www.cemexventures.com/most-sustainable-cities-world/
44. Jersery Girl Organics, 10 Sustainable Farming Practices and Why they Matter. 8 Nov 2021. https://www.jerseygirlorganics.co.nz/post/10-sustainable-farming-practices-and-why-they-matter
45. F. Sandison, J. Yeluripati, D. Stewart, Does Green Vertical Farming Offer a Sustainable Alternative to Conventional Methods of Production?: A Case Study from Scotland. Food Energy Secur. **12**(e438) (2023)
46. J.K. Hawes, B.P. Goldstein, J.P. Newell, E. Dorr, S. Caputo, R. Fox-Kämper, B. Grard, R.T. Ilieva, A. Fargue-Lelièvre, L. Poniży, V. Schoen, K. Specht, Comparing the Carbon Footprints of Urban and Conventional Agriculture. Nat. Cities (2024)
47. IPCC, International Panel on Climate Change, Global Warming of 1.5°C.An IPCC Special Report on the Impacts of Global Warming of 1.5°C Above Pre-industrial Levels and Related Global Greenhouse Gas Emission Pathways, in the Context of Strengthening the Global Response to the Threat of Climate Change (2018)
48. ISO 14068–1:2023, *Climate Change Management—Transition To Net Zero—Part 1: Carbon Neutrality* (Vernier, Geneva, 2023)

Chapter 7
Sustainable Consumption and Use

Abstract It may be argued that, by definition, consumption is unsustainable, especially considering current structures and current levels of consumption. As long as satiation is not in sight, "sustainable consumption" is an oxymoron. However, structural changes that make the "right choice the easy choice" and the use of education/marketing as an agent of change have the potential to reduce the inherent contradiction of "sustainable consumptions". This chapter describes different approaches to support this change, focusing on service approaches, product improvements, communication, and marketing as well as regulatory aspects.

Keywords Sufficiency · Basic needs · Services · Durability · Rebound effects · Greenwashing · Regulations · Reporting · Consumer milieus

It can be argued that consumption is unsustainable by definition at least with current structures and at current levels where there is no satiation in sight, i.e. "sustainable consumption" is an oxymoron. However, structural changes that make the "right choice the easy choice" [1] together with education/marketing as an agent of change have the potential to reduce the inherent contradiction of this oxymoron.

Consumers can get information about the carbon or water footprint of products. There are even websites allowing the calculation of the individual carbon footprint, for example [2–4]. Consumers can quantify the effect of changes to their personal consumption habits by entering different values for different consumption patterns. Including life cycle thinking into daily consumption decisions is a responsible way to more closely align the personal behavior with sustainability. In general, impacts are lower for vegetarians/vegans and those who do not use planes or cars. A lot of other general rules can be derived from these calculators.

In general, higher income leads to higher consumption levels that drive these footprints. The extent of negative environmental impacts of consumption is mainly determined by the wealth of consumers. The higher the income, the further sustainability limits are exceeded. The "overconsuming wealthy and middle classes" have

© The Author(s), under exclusive license to Springer Nature Switzerland AG 2024

W. Schmidt, *Solutions For Sustainability Challenges*,
https://doi.org/10.1007/978-3-031-63624-0_7

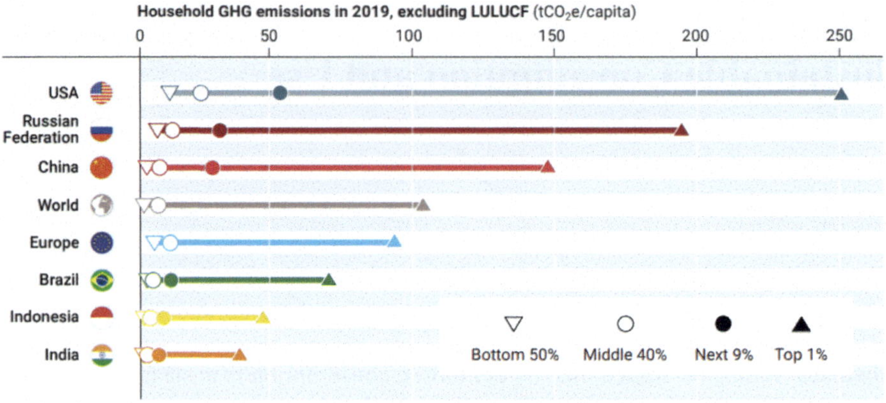

Fig. 7.1 Per capita greenhouse gas emissions per country and income [8] *Note* Per capita emissions include emissions from domestic consumption, public and private investments, and imports and exports of carbon embedded in trade with rest of the world. Households are ranked according to total emissions and divided accordingly into groups (e.g. the bottom 50% refers to the 50% of households with the lowest emissions in that country or region). *Source* Chancel et al. (2022)

been identified as a major issue due to their "wasteful consumption", "environmentally malign technological choices such as private automobiles, plastic everything" and "growthmania" [5].

Of course, exemptions from the rule exist, however, these exemptions are rare and require significant strength of character as well as discipline to ensure consistency, especially in non-supportive societal environments, over the individual's life time.

Globally, the 10% of the population with the highest income accounts for nearly half (48%) of global emissions. Two-thirds of this group live in developed countries. The bottom 50% of the world population contribute only 12% of total emissions [6]. The upper 50% of the world population (high- and upper-middle-income countries) are responsible for 90% of air travel emissions, while lower-middle-income countries emit just 1% [7] (Fig. 7.1).

Usually, the negative environmental impact of consumption increases with the wealth of individuals. Each person has the responsibility to check individual impacts & change accordingly.

7.1 Less Is More

Sustainable sufficiency is the most efficient and effective approach to adjust the individual, societal, and/or global footprint but it is not popular in most countries and political systems. Sufficiency targets a central root cause for human exceedance of sustainability limits. The idea is to scale back demand to basic needs. The term "basic needs" seems easy to define. The "Sustainable Society Index" only covers sufficient food, sufficient to drink, and safe sanitation as basic needs [9]. However, the perception of what should be part of these "basic needs" is of course subject to individual, societal, and cultural influences and changes over an individual's life time and with opportunities of supply. Considering that the definition of sustainable development—meeting the needs of the present without compromising the needs of future generations—is based on the concept of "need", it is important to define the meaning of the needs to be met. "Needs" for sustainable development refer to the "essential needs of the world's poor, to which overriding priority should be given" [10] rather than middle- or upper-class incremental needs that exceed basic needs.

If we take the example of private vehicles and only look at the carbon emissions of a vehicle from raw material extraction to material, part, and vehicle production (cradle-to-gate), there is a clear positive linear relation between the weight (and size) of the vehicle and the resulting carbon emissions regardless of design (Fig. 7.2). This assessment does not take the use of the vehicle into account where heavier, larger vehicles frequently perform worse due to their worse aerodynamics. Considering that most of these vehicles offer 5 seats (apart from the two heaviest vehicles in Fig. 7.2), the demand for heavier, taller, and bigger vehicles is not a sign of sustainable responsibility. Individual reasons may exist why a larger trunk volume and higher, easier accessible vehicles are seen as necessary. However, in most cases it is difficult to rationalize this as a basic need that cannot be addressed differently.

However, only using smaller vehicles is not sufficient either to get to carbon neutrality. Ideally, all material usage needs to be scrutinized and a "cultural shift to prioritize immaterial ways to fulfill needs, and invest in health, well-being, education and decent jobs, rather than material accumulation" is needed [7].

It is important to realize that usually the very effective approach of sufficiency is systematically excluded from most (life cycle) discussions by the introduction of reference units and "functional units" (see Sect. 2.2.1.1) [11]. The focus on relative impacts enables the comparison of alternative options in terms of their ability to satisfy the undisputed demand as expressed by the functional unit. Yet, it does not allow the comparison of alternatives that do not satisfy the same level of function or demand. Therefore, there are proposals to include sufficiency needs in the assessment of reference values [12]. For example, in the building sector, the use of area-related reference values like impact per square meter does not account for opportunities of area reduction. The same applies to all basic needs and more fundamentally for needs that are not basic needs where their absence is the most effective sustainable option. In the previous example of vehicles this would mean that the functional unit is too narrowly defined by requiring a certain vehicle segment or mobility mode.

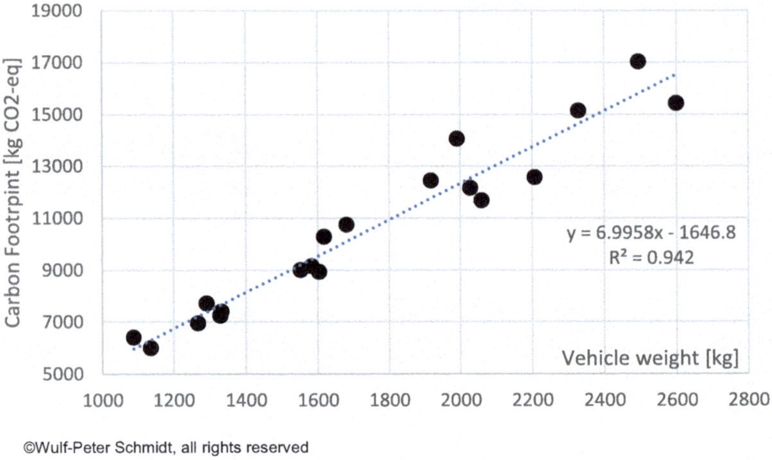

Fig. 7.2 Linear relation between vehicle weight and cradle-to-gate carbon emissions in early 2020's (own calculations). ©Wulf-Peter Schmidt, all rights reserved

> The sustainability definition of "meeting the needs" refers to basic needs of the world's poor. Beyond this, sufficiency is the most effective sustainability option although it is excluded in Life Cycle Tools.

New products and services combined with effective marketing can change the definition of basic needs. A good example is digitalization—one could speculate that human rights would have included free internet access and smartphones alongside with shelter and food as basic needs if already available centuries ago.

The increased availability of goods has led to non-sustainable consumption levels that trigger the exceedance of sustainable limits also due to marketing patterns raising demand. However, there is sufficient research outlining that marketing patterns for sustainability can be positive. A literature review of different sustainable marketing approaches [13] concluded that consumers need intrinsic motivators for a more sustainable consumption. This is best based on a "non-materialistic mindfulness mindset".

Besides technology innovation and availability, other influences on sufficiency also exist, including culture, peer groups, role models, advertisement, and many more.

7.2 Sharing and Pay-Per-Service

If products can be used multiple times by multiple people, another way of improving the sustainability footprint is to change the ownership models of products and share them. This reduces the sustainability footprint of the used product and is most effective for production-sensitive (Sect. 2.1.1), end-of-life-sensitive (Sect. 2.1.3), and transportation-sensitive (Sect. 2.1.4) products. The impact for usage sensitive products (Sect. 2.1.1) is limited unless the usage is indifferent to the number of individuals benefitting from it at the same time (e.g. same light or heat in the same room).

Sharing platforms as well as Pay-per-service schemes help customers to undergo a cultural change from ownership-centric to utility-centric perspectives. This can increase the incentive to reduce consumption levels when compared to a situation where, for instance, an owned, underutilized car is waiting anyways in front of the own home, i.e. it is easier to use cars instead of other mobility options. On the other hand, such schemes may lead to additional consumption because the entry-cost is reduced—e.g. for a (second or third) car.

Another ambivalent example is shared e-scooters that in some studies lead to 50% more trips [14] while predominantly replacing walking trips (30–60%) or public transport (wide range from 3% in areas with low public transport options to 34% in areas with high availability although e-scooter are often helpful in integrated public transportation) rather than supporting a modal shift away from cars [15]. Also, shared e-scooters are used less for commuting than (shared) bicycles. The absent modal shift away from cars is also the result of use patterns in Europe (e-scooters are used by younger people in cities that have a lower motorization share). In the US, where cars are generally used for much shorter distances than in Europe and society mainly relies on cars, e-scooters have a bigger impact on the modal shift away from cars. The reason is that 30.8% of car trips are below 2 miles—a typical e-scooter trip range [15]. Where e-scooter trips are incremental or replace walking or bicycles, the environmental and safety impact of e-scooters is always negative. For other cases, the comparison depends on further details. For example, the life cycle performance of e-scooters is significantly impacted by the way e-scooters are collected for recharging and re-distributed as well as by their lifetime. On average, e-scooters have higher life cycle carbon emissions than bicycles and public transport; if car trips are replaced, they are advantageous if replacing internal combustion engines [16].

A positive pay-per-service example is “ColdHubs” in Nigeria where solar-powered cooling facilities can be used for freshly harvested food on a more affordable pay-as-you-store subscription model that avoids food spoilage and reduces CO_2 emissions due to more efficient technology [7].

The efficacy of sharing models depends on the sector (best for production-sensitive products), the impact on use intensity and the alternatives that are replaced in the specific context.

7.3 Extending Product Life

The concept of "slower consumption" [17] is especially relevant for production-sensitive products. It tries to extend the effective product life span by careful use and maintenance as well as second usages. From an environmental life cycle perspective, the optimal life span can be calculated by comparing it to alternatives like substitution with a new but lower-impact product. The smaller the use phase advantage of a new lower-impact product compared to the current product is, the more it makes sense to extend the useful product life [18].

Extending product life or slowing down the "throughput" society is undermined by non-repairability—by design or because of high labor costs or the required repair expertise (for example for electronics). Another barrier is the functional obsolescence including missing backwards compatibility/missing support for previous software/ electronic levels, and the power of marketing [19]. Regulators like the EU are trying to counter this trend with a "right for repair". In reality, extending the product life may require more than physical/functional aspects. Factors like timeless aesthetics and continued meaningfulness of the product for consumers as addressed in the Emotionally Durable Design (see Sect. 4.2) must be considered as well.

Extending the product life helps sustainability more the higher the negative impacts of production and the smaller the advantage of new products are during use; products must stay relevant for consumers.

7.4 Addressing Rebound Effects

Higher efficiency of (green) production can lead to lower prices allowing people to consume more—which may (partly) neutralize the initial savings by leading to the so-called "rebound effect". It may even lead to "Backfire effects" where the additional consumption as a result of the saved money from higher efficiency results in even higher emissions than the saved emissions by the more efficient product. One can also argue that developmental aid can create rebound/backfire effects if the development aid does not steer development in more sustainable directions. However, rebound forecasts are difficult. Disruptive changes in behavior, demand, and products cloud the prediction capabilities.

Cheaper green electricity and lower-cost EVs generate more demand and therefore more production—also because customers might be less ashamed of their consumption patterns. On the other hand, in such a scenario, products might get irrelevant like cars for some citizens in big cities with well-established public transport options.

Sufficiency is a potential way out. Some suggest abolishing capitalism; on a national level Bhutan attempted this together with introducing the right for happiness—and has reportedly failed so far [20]. Therefore, the most constructive way seems to invest the saved money in sustainable start-ups and assets (see Sect. 9.2.4) because it helps the required transformation towards sustainability.

Rebound / backfire effects reflect the impact of additional consumption enabled by lower costs of more efficient (green) products. Therefore, beyond technical measures behavioral changes are important.

7.5 Life Cycle Thinking in Consumption—Sustainable Actions by Everybody

Life cycle thinking in consumption means that private and business consumption decisions are made on the basis of a thoughtful process that balances the life cycle implications of the own consumption patterns on sustainability. The idea is that informed and responsible consumers have a big lever for sustainability. The individual and combined demand signal steers the market away from unsustainable products, services, and production, i.e. uses the existing demand-supply dynamics to address sustainability. Besides this indirect influence, the positive direct impact can be significant—as demonstrated by crowd investing, crowd intelligence, etc. where each contribution counts and helps. Every company—in tourism, agriculture and the food industry, consumer goods industry as well as all industry, transport, and mobility organizations linked to them—only do things for which they hope to find sufficient paying customers. If there is insufficient demand for flights to any destination, for non-seasonal food like strawberries in winter time, for meat, or for any other product none of these will be offered. Likewise, if there is no demand for more sustainable products, these will not be offered either.

The two main conditions to allow this to work are

(a) The availability and comprehensibility of information about the life cycle implications of consumer decisions.
(b) Willingness, ability, and prioritization on the consumer side to act accordingly.

Several regulations aim to improve the availability of information about sustainability including label regulations (EU Ecolabel, Energy and CO2 labels, etc. [21] or the ISO 14020 series) and transparency regulations (see for example Sect. 5.1). The Product Environmental Footprint (PEF) or Organization Environmental Footprint (OEF) as environmental LCA methods are the basis for environmental life cycle-based information requirements in the EU. In the public domain, consumer guidance is provided by environmental credentials of products but this is rarely done from a life cycle perspective (e.g. Green NCAP moves in that direction for vehicles [22]). Only a few regulatory activities exist to ensure transparency of all material sustainability aspects including social issues (e.g. Taxonomy, Corporate Sustainability Reporting Directive). The intelligibility of the information is often inversely proportional to the completeness and extent of transparency: the perfect is the enemy of the good. Well-intentioned approaches like making LCAs/PEFs available to consumers conflict with the time consumers have to make a decision as well as the capacity to deal with the complexity of LCAs or risk that trade-offs or uncertainties lead to the conclusion/excuse that it does not matter.

Thus, simplifications to single-scores or single dimensions of sustainability are often preferred although scientifically questionable. Most common are tables showing the carbon footprint of products and services ([23–27] and many more).

The main conclusions of these examples are that for any food-related decision, meat (in particular beef) and dairy products should be avoided (i.e. vegetarian or even vegan products are best, even if vegans eat non local, highly processed plant-based foods). For mobility, the message is that lower energy consumption and shorter distances are better, i.e. walking is better than bicycles that are better than busses/trains that are better than cars (the smaller and the more electrified and the more people/goods transported the better) with flying being the worst option. Similarly, technological products' carbon footprints are dependent on their utilization rate, size, and energy consumption.

Consumers' willingness to fundamentally change current consumer behaviors and to prioritize sustainability can be addressed by

- Education/appeals or by financial incentives for better options.
- Financial disincentives for worse options as well as prohibitions or bans of the worst products or substances.

The most successful approach is to make it uncomfortable or embarrassing to follow non-sustainable consumption patterns. Examples include prohibitive taxation (more effective than offering incentives as proven for several examples such as vehicles, food, etc.) and increase time efforts for non-sustainable solutions (e.g. by removing parking spaces or road lanes for vehicles). Again, the perfect is the enemy of the good—in this case the pressure needs to be increased carefully to avoid defiance and political backlashes.

Life Cycle considerations can form the basis for individual consumption decisions based on information offered; willingness is best increased by making non-sustainable choices less comfortable.

7.6 Sustainable Services

There are several services that inherently and predominantly contribute actively towards sustainability.

Healthcare, social care, and medical services by definition contribute positively to social sustainability, human rights and the UN Sustainable Development Goal (SDG) 3 "Good health and well-being". The specific 2030 targets include the reduction of human mortality, ending specific epidemics, reducing non-communicable diseases and promoting mental health, improving access to health services, etc. This SDG is also linked to environmental (reducing mortality and illnesses from pollution) and economic aspects (increase health financing) as well as innovation (vaccine and medicine research).

Training and educational services directly contribute to UN SDG 4 "Quality education" that demands equal access to inclusive, free, or affordable high-quality education. From a sustainability perspective, one sub-target for 2030 is to "ensure that all learners acquire the knowledge and skills needed to promote sustainable development, including, among others, through education for sustainable development and sustainable lifestyles, human rights, gender equality, promotion of a culture of peace and non-violence, global citizenship and appreciation of cultural diversity and of culture's contribution to sustainable development".

Repair, refurbish, maintenance, and remanufacturing services contribute to a circular economy and therefore actively contribute to a sub-target of UN SDG 12 "Responsible consumption and production". These services lead to a slow down of material flows, improve environmental performances, and typically have socio-economic advantages including more affordable customer prices and creating jobs.

Medical, health care and educational as well as repair, refurbish and maintenance services commonly automatically contribute to sustainable development.

7.7 Sustainable Marketing and Communications

Sustainability in marketing typically consists of avoiding misleading greenwashing statements. More and more regulations, standards, and rules are being implemented to restrict vague claims that are open to interpretation. These include terms such as "green", "environmentally friendly" and "sustainable". "Net zero", "carbon neutral" or even "carbon positive" are also terms subject to clear rules. Sustainable marketing and communication can have constructive and positive impacts on sustainability if they advertise for more sustainable consumption patterns. They should be based on a longer-term relationship between consumers and producers, fostering an understanding of consumer needs in the context of socio-environmental transformations.

7.7.1 *Sustainability Reporting*

The number, extent, and stringency of regulations for reporting sustainability has recently increased significantly. This is especially the case in Europe and the US, but also globally through the introduction of the International Sustainability Standards Board (ISSB). The goal is to ensure that sustainability claims are substantiated, consider the entire life cycle, and are verified to ensure they are reliable sources of information for investors as well as customers. This is similar to the internationally standardized financial reporting systems.

For investors, integrated reporting systems have been established that reflect on performance in terms of financial success and sustainability. Over time, new regulations will elevate sustainability reporting to the level of financial reporting in terms of the extent and granularity required but also the accuracy and rigor of information. So far, the EU Corporate Sustainability Reporting Directive (CSRD) [28] is the most far-reaching reporting regulation—with other regions planning to follow (e.g. China). It requires the disclosure of more than 1000 data points regarding Environment, Social and Governance (ESG) criteria as detailed in the European Sustainability Reporting Standards (ESRS) [29]. This will be implemented—depending on the size and region of the organization—for the fiscal year 2024 and covers a wide range of aspects in the areas of:

- Cross-cutting aspects—like a materiality analysis (see Sect. 3.2; although some information needs to be disclosed regardless of materiality), strategy, impact risk, and opportunity management.
- Environment—Climate, Pollution, Water and Marine Resources, Biodiversity and Ecosystems, Resource use and Circular Economy.
- Social—Own Workforce, Workers in the Value Chain, Affected Communities, Consumers and End-users.
- Governance—Business Conduct.

The European Financial Reporting Advisory Group (EFRAG) initially drafted the ESRS to be valid across the EU while the CSRD will be or has already been transposed into the national law of EU member states.

One requirement of the CSRD is that absolute carbon reduction targets must be set. It must also be explained whether this is a "science-based target". The report also has to compare the chosen target to the change required for a 1.5 °C pathway if the chosen carbon reduction target does not align with a 1.5 °C pathway. Similarly, biodiversity targets must include milestones and inform on the alignment with relevant frameworks that aim for "no net loss", "net gain" or "full recovery". These are just two examples that convey that the intention of the CSRD extends beyond organizations disclosing information. It also aims to drive changes in organizations and their strategies as they may want to avoid admitting that they have no or insufficient commitments or strategies in place to achieve internationally agreed climate change targets. These disclosure requirements enable investors and commercial customers to obtain sufficient information for their investment and supply decisions.

Although this is an EU regulation, CSRD also applies to companies of a certain size or groups outside the EU and affects all companies with securities listed on an EU-regulated market or EU stock exchanges. For smaller EU companies, transitional and/or simplified reporting standards may apply.

The CSRD is the successor to the Non-Financial Reporting Directive (NFRD) that "only" applied to around 11,700 businesses who were required to report on their business model, policies related to due diligence, outcome of these policies, main risks linked to the companies' operations, and other key non-financial performance indicators.

CSRD sustainability reporting encourages companies to further improve on their strategy & performance and provides transparency for sustainable investment & purchasing decisions.

The EU Sustainable Finance Disclosure Regulation (SFDR) firstly requires disclosures on an entity level regarding potential negative impacts of investment decisions on ESG factors, whether such ESG risks are considered in the decision-making process and how management remuneration align with the integration of sustainability risks. Secondly, reporting on a fund/product-level is required. This report must detail how sustainability risks might impact financial performance, whether and how the product considers potential negative impacts on sustainability, and how products labeled as sustainable investments monitor, measure, and assess their sustainability impact. The SFDR is linked to the EU Taxonomy Regulation (see Sect. 9.2.1) and applies to financial market participants and financial advisors. It aims to improve transparency about the inclusion of sustainability risks in investment decision-making processes of financial products that either promote environmental characteristics or claim to be sustainable investments.

For climate change, specific regulations like the UK Climate-related Financial Disclosure Regulations (TCFD) exist as well. The link between ESG and financial reporting is also referred to in the International Sustainability Standards Board (ISSB). In March 2022, the ISSB published the "'Exposure Draft IFRS S1 General Requirements for Disclosure of Sustainability-related Financial Information" in which it proposed requirements for entities to disclose financial information related to risks to and opportunities for sustainability.

The EU Directive "on substantiation and communication of explicit environmental claims (Green Claims Directive)" [30] aims to ensure that environmental claims are substantiated by requiring entities to "demonstrate that environmental impacts, environmental aspects or environmental performance that are subject to the claim are significant from a life cycle perspective". By linking reporting requirements to life cycle thinking and tools, the directive reduces misleading claims. Through this directive, the EU also restricts the use of terms like "carbon neutrality" if it is based on offsetting.

Similarly, the EU Taxonomy regulation [31] requires the disclosure of key performance indicators (KPIs) relating to six environmental objectives to allow investors to understand the extent to which assets or incomes are generated by sustainable practices.

In the US, sustainability disclosure regulations around the use of conflict minerals exist. Section 1502 of U.S. Dodd Frank Act requires companies listed in the US to disclose whether they use "conflict minerals" (tin, tungsten, tantalum, and gold) and whether these minerals originate in the Democratic Republic of the Congo (DRC) or an adjoining country. This is a narrower definition (both in terms of types of materials as well as sourcing countries) than the one adopted in European due diligence regulations (see Sect. 5.1). The Uyghur Forced Labor Prevention Act (H.R. 6256) is a federal law that addresses reported forced labor practices in China's Xinjiang Uyghur Autonomous Region. The intention is to prevent American entities from funding forced labor of ethnic minorities in the region. In addition, on a state level, the California Senate Bill 253 requires emission disclosures for large corporations operating in California with a total annual revenue equal to or above $1bn by 2026 (for scope 1 and 2) or 2027 (for scope 3).

ISO 14068:2023 [32] attempts to standardize the communication of carbon neutrality activities. It includes information about the development, determination, and calculation of carbon neutrality and requires transparency regarding potential adverse effects on the environment and society. Any communication on carbon neutrality should be transparent, conservative in all assumptions, values, and procedures but also science-based, ambitious, and recognize the urgent need for action.

All regulations and standards seek to avoid "greenwashing" which can be associated with any of eight criteria ("sins"): hidden trade-offs, lack of evidence, vagueness, irrelevance, distraction (by pointing to the lesser of two evils), deception [33], lack of ambition, self-regulated or pseudo-labels. The concern about greenwashing is not only that it misleads consumers, but also that false claims about sustainability undermine the competitive edge of genuinely more sustainable products or companies. The risk of "greenhushing", where organizations stop making any sustainability claims to avoid legal issues if they cannot prove or meet their promises, exists. Another reason for greenhushing might be an attempt to avoid ideological rejections if the target groups are not open to sustainability or ESG messages.

7.7.2 *Marketing Impact*

Marketing is about the promotion and sales of products and services. It anticipates and satisfies the needs and wants of the buyer and aims to turn a profit. Marketing includes all activities that bring buyers and sellers together and can be defined as "the process of developing, pricing, promoting and distributing products, services or ideas that are tailored to the market; it includes all other activities that create value and systematically lead to increased sales or another desired response, establish a good reputation and ongoing relationships with customers, so that all stakeholders achieve their objectives" [34].

According to this definition, marketing directly impacts the levels and sustainability of consumption. The intention is to increase the sales of products and services. Therefore, marketing inherently conflicts with the "less is more" approach of sustainable consumption (see Sect. 7.1). In this context, "macro-marketing wants to address and integrate concerns about the social and environmental impacts of marketing activity and the relationships between markets, regulation, and social welfare. It tried to systematically consider the inter-relationship between marketing and society with an emphasis on the potential impacts on environmental quality and societal welfare" [35]. Classic marketing tools such as pricing, product, distribution, and communication can enhance the sustainability of an organization:

- Pricing can discourage less sustainable products if external costs are internalized (see Sect. 2.3.2). It can also encourage the purchase of more sustainable products by offering discounts for all customers or specific social groups. Pricing can also ensure better remuneration along the value chain.
- Product development benefits from clear input from marketing departments. Marketing departments should communicate the sustainability needs, expectations, and trends of current and future markets and customer groups internally.
- Choosing specific distribution channels can minimize social and environmental impacts along the value chain (see Sect. 5.4).
- Communication can guide customers towards more sustainable consumption patterns.

Different attempts have been made to use marketing in a more positive way by trying to advocate for more responsible consumption patterns and enhance customers' willingness to consume sustainably. This is often supported by purported improvements in products and fair-trade labels, ecolabels, etc. However, some still see these efforts as green washing or, in the best case, superficial improvements.

In addition, ethical or green marketing activities still use persuasive approaches. They try to impact customers and focus on selling products and services instead of trying to work together with customers in a deeper relationship that extends beyond the specific product or service. Therefore, a sustainability marketing approach should build and maintain sustainable relationships with customers as well as the social and natural environment. It supersedes classical marketing that focusses on creating a product transaction (gaining new customers) or relationship marketing that focuses on long-term relationships with customers and suppliers to add value for all involved parties by collaboration. Sustainability marketing includes the long-term orientation of relationship marketing as opposed to the conventional short-term transaction focus of modern marketing. Additionally, it looks beyond economic market aspects and also considers societal and environmental sustainability. Sustainability marketing is based on four key areas [35, 36]:

- Treating sustainability challenges as the starting point of marketing processes and not as a set of externalities or constraints.
- Understanding consumer behaviors holistically and extending the classical market research by an analysis of the socio-environmental priorities.
- Reconfiguring the marketing mix by integrating social, ethical, and environmental values to replace a producer and product-focused approach.
- Appreciating and utilizing the transformational potential of marketing activities and relationships based on an analysis of how marketing can contribute to a transformation to a more sustainable society.

Marketing must change from a short-term focus on sales to a lasting relationship that creates an understanding of consumer needs in the context of socio-environmental transformation needs.

Without this, "marketing will continue to drive global crises linked to climate change, poverty, food shortages, oil depletion and species extinction, instead of helping to tackle them" [36] (Fig. 7.3).

To understand customer needs and values better, customers are sometimes categorized into discrete groups. One example is the so-called sinus milieus, in which people are aggregated into groups with similar values and sensibilities as well as life goals and lifestyles. Sinus milieus have two dimensions: basic beliefs and socio-economic realities. Marketing and communications can target these milieus through choosing specific topics, types of approach, and touch points. These milieus include the LOHAS (Lifestyle of Health and Sustainability) or the LOVOS (Lifestyle of Voluntary Simplicity) also called Socially Conscious Upwardly-Mobile People (SCUPPIE). LOVOS follows the "less is more" approach (see Sect. 7.1) [37].

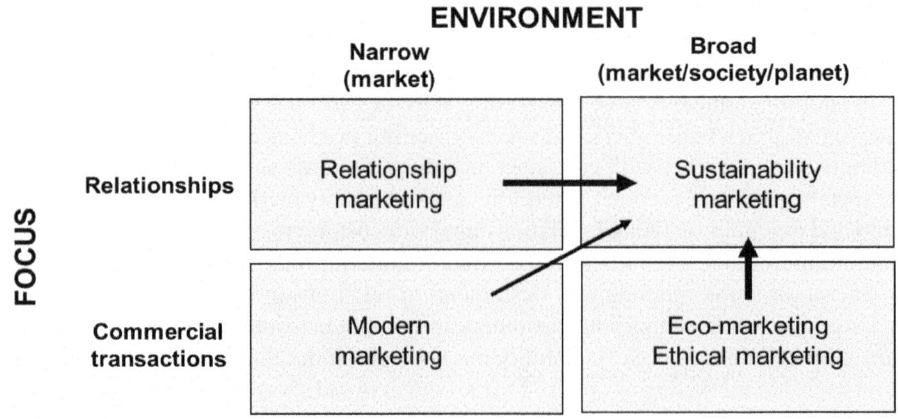

Fig. 7.3 Different types of marketing [35]

For all consumer groups, the credibility of sustainability messages is important but for LOHAS and LOVOS/SCUPPIE, the product design, appearance, style, and functionality must support this. In addition, more product details must be shared—including product life cycle information.

Sustainable marketing does not advertise any product as sustainable. Instead, sustainable marketing helps potential consumers meet their needs in the most sustainable and informed way possible. This is done by:

- Thorough identification and analysis of people's needs and values based on long-term perspectives and relationships.
- Transparent, comprehensive /understandable, relevant, validated, clear and specific information that does not conceal trade-offs, is not misleading or deceptive, and does not distract from other, more important negative sustainability impacts (the so-called "sin of the lesser of two evils").
- Providing guidance and feedback to the own organization's product development and other relevant departments regarding meeting customer needs more sustainably.

Sustainable marketing is not about advertising products as sustainable. Rather, it facilitates meeting needs in the most sustainable way by appropriate internal & external information.

References

1. J. Thøgersen, Unsustainable Consumption—Basic Causes and Implications for Policy. Eur. Psychol. **19**(2) (June 2014)
2. WWF, How Big Is Your Environmental Footprint? [Online], https://footprint.wwf.org.uk/#/. Accessed 5 Mar 2023
3. UBA, UBA Carbon Calculator [Online], https://uba.co2-rechner.de/en_GB/. Accessed 5 Mar 2023
4. Waterfootprint, Personal Calculator—Extended, Water Footprint Network [Online], https://waterfootprint.org/en/resources/interactive-tools/personal-water-footprint-calculator/personal-calculator-extended/
5. R. Dirzo, G. Ceballos, P.R. Ehrlich, Circling the Drain: The Extinction Crisis and the Future of Humanity. Philos. Trans. **377**(1857) (15 August 2022)
6. UNEP, *Broken Record—Temperatures Hit New Highs, Yet World Fails to Cut Emissions (Again)* (United Nations Environment Programme, Nairobi, 2023)
7. Circle Economy, *The Circularity Gap Report 2023,* Amsterdam, 2023
8. United Nations Environment Programme (UNEP), *Emissions Gap Report 2022: The Closing Window—Climate Crisis Calls for Rapid Transformation of Societies* (UN, Nairobi, 2022)
9. M. Saisana, D. Philippas, *Sustainable Society Index (SSI): Taking Societies' Pulse Along Social, Environmental and Economic Issues* (EU Commission—Joint Research Center, Ispra, Italy, 2012)

10. United Nations, *Our Common Future (The Brundtland Report): World Commission on Environment and Development* (UN, 1987)
11. ISO14040, *Environmental Management—Life Cycle Assessment—Principles and Framework* (International Organization for Standardization, 2006)
12. A. Hock, J. Jäger, A. Rietz, Sufficiency as a Criterion for Sustainability. Earth Environ. Sci., 1078–1087 (2022)
13. M. Haider, R. Shannon, G.P. Moschis, Sustainable Consumption Research and the Role of Marketing: A Review of the Literature (1976–2021). Sustainability **14**(6), 28 (March 2022)
14. R.L. Sanders, V.d.S. Brum-Bastos, T.A. Nelson, Insights from a Pilot Investigating the Impacts of Shared E-Scooter Use on Physical Activity Using a Single-Case Design Methodology. J. Transp. Health **25**, 101379 (2022)
15. K. Wang, X. Qian, D.T. Fitch, Y. Lee, J. Malik, G. Circella, What Travel Modes Do Shared E-Scooters Displace? A Review of Recent Research Findings. Transp. Rev. **43**(1), 5–31 (2023)
16. J. Hollingsworth, B. Copeland, J.X. Johnson, Are E-Scooters Polluters? The Environmental Impacts of Shared Dockless Electric Scooter. Environ. Res. Lett. **14**, 084031 (2019)
17. T. Cooper, Slower Consumption—Reflections on Product Life Spans and the "Throwaway Society." J. Ind. Ecol. **9**(1–2), 51–67 (2005)
18. H. Kunst, *Ökologische Optimierung von Substitutionsentscheidungen* (Dissertation.de, Berlin, 2003), p. 178
19. S. Suppipat, A.H. Hu, A Scoping Review of Design for Circularity in the Electrical and Electronics Industry. Resour. Conserv. Recycl. Adv. **13**, 2667–3789 (2022)
20. F. Schätzing, *Was, wenn wir einfach die Welt retten? Handeln in der Klimakrise* (Kiepenheuer & Witsch, Köln, 2022)
21. EU Commission, Products—Labelling Rules and Requirements—Rules and Requirements for Labelling Eco-Friendly Products, Energy, Fuel Consumption and Chemicals [Online], https://commission.europa.eu/energy-climate-change-environment/standards-tools-and-labels/products-labelling-rules-and-requirements_en.
22. Green NCAP, https://www.greenncap.com/, Euro NCAP [Online], https://www.greenncap.com/.
23. H. Ritchie, *You Want to Reduce the Carbon Footprint of Your Food? Focus on What You Eat, Not Whether Your Food Is Local* (Our World in Data, 2020), https://ourworldindata.org/food-choice-vs-eating-local.
24. Carbon Footprint of Common Consumer Products. Eupedia [Online], https://www.eupedia.com/ecology/carbon_footprint_consumer_products.shtml. Accessed 25 Nov 2023.
25. C.J. Meinrenken, D. Chen, R.A. Esparza, V. Iyer, S.P. Paridis, A. Prasad, E. Whillas, The Carbon Catalogue, Carbon Footprints of 866 Commercial Products from 8 Industry Sectors and 5 Continents. Scientific Data—Article (2022)
26. Tesco, Product Carbon Footprint Summary (2012)
27. A. Kustar, D. Patino-Echeverri, A Review of Environmental Life Cycle Assessments of Diets: Plant-Based Solutions Are Truly Sustainable, Even in the Form of Fast Foods. Sustainability **17**(13), 9926 (2021)
28. European Union, DIRECTIVE (EU) 2022/2464 OF THE EUROPEAN PARLIAMENT AND OF THE COUNCIL of 14 December 2022 Amending Regulation (EU) No 537/2014, Directive 2004/109/EC, Directive 2006/43/EC and Directive 2013/34/EU, as Regards Corporate Sustainability Reporting (Brussels, 2022)
29. European Commission—Finance [Online], https://finance.ec.europa.eu/regulation-and-supervision/financial-services-legislation/implementing-and-delegated-acts/corporate-sustainability-reporting-directive_en.
30. EU Commission, *Proposal for a DIRECTIVE OF THE EUROPEAN PARLIAMENT AND OF THE COUNCIL on Substantiation and Communication of Explicit Environmental Claims (Green Claims Directive)* (EU, Brussels, 2023)
31. EU Commission, Regulation (EU) 2020/852 of the European Parliament and of the Council of 18 June 2020 on the Establishment of a Framework to Facilitate Sustainable Investment, and Amending Regulation (EU) 2019/2088 (Brussles, 2020)

32. ISO 14068–1:2023, *Climate Change Management—Transition to Net Zero—Part 1: Carbon Neutrality* (Geneva, Vernier, 2023)
33. S. Dimitrieska, A. Stankovska, T. Efremova, The Six Sins of Greenwashing. Econ. Manag. **13**(2), 82–89 (2017)
34. B. Verhage, *Marketing Fundamentals* (Noordhoff Uitgevers, Groningen/Houten, The Netherlands, 2013)
35. K. Peattie, F.-M. Belz, Sustainability Marketing—An Innovative Conception of Marketing. Thexis **27**, 8–15 (2010)
36. F.-M. Belz, K. Peattie, *Sustainability Marketing: A Global Perspective* (Frank-Martin Belz, 2023)
37. F. Müller-Friemauth, N. Klein-Reesink, C. Harrach, *LOHAS: Mehr als Green-Glamour - Eine soziokulturelle Segmentierung* (Sinus Sociovision, KarmaKonsum, Heidelbert, Frankfurt a.M., 2009)

Chapter 8
Sustainable End-of-Life Operations

Abstract End-of-life operations cover all operations that support a circular economy or waste disposal. They include collection, pre-processing, re-use, repair, refurbishing, remanufacturing, recycling, energy recovery, measures to reduce or remove any negative impacts of (hazardous) waste as well as processes that support biological recovery processes like composting. Final disposal is the alternative to circular economy. Sustainable end-of-life operations optimize the environmental, economic and social value of products and services. The overall aim of sustainable end-of-life operations is to maintain material flows in the useable technosphere for as long as possible, maximize the environmental, social and economic benefits of these and minimize any negative implications. This chapter demonstrates the organization of material flows in various circularity approaches within or between product life cycles to allow efficient material utilization cascades. Natural materials, fiber reinforced polymers, EV batteries and rare earth containing magnets illustrate these principles.

Keywords Circular economy · Recycling · Reuse · Remanufacturing · Recycled materials · Cleaning · Comminution · Classification · Sorting · Separating · Composites · Fiber reinforced polymers · Batteries · Rare earth

End-of-life operations cover all operations that support a circular economy or final disposal. They include collection, preprocessing, reuse, repair, remanufacturing, recycling, energy recovery, measures to reduce or delete any negative impacts of (hazardous) waste as well as processes like composting to support biological recovery processes [1]. Final disposal is the alternative end-of-life operation. Sustainable end-of-life operations are carried out to optimize the environmental, economic, and social value.

The overall aim of sustainable end-of-life operations is to keep material flows in the useable technosphere as long as possible, maximize the environmental, social, and economic benefits of this approach and minimize any negative implications. Figure 8.1 shows the organization of material flows based on different circularity

© The Author(s), under exclusive license to Springer Nature Switzerland AG 2024
W. Schmidt, *Solutions For Sustainability Challenges*,
https://doi.org/10.1007/978-3-031-63624-0_8

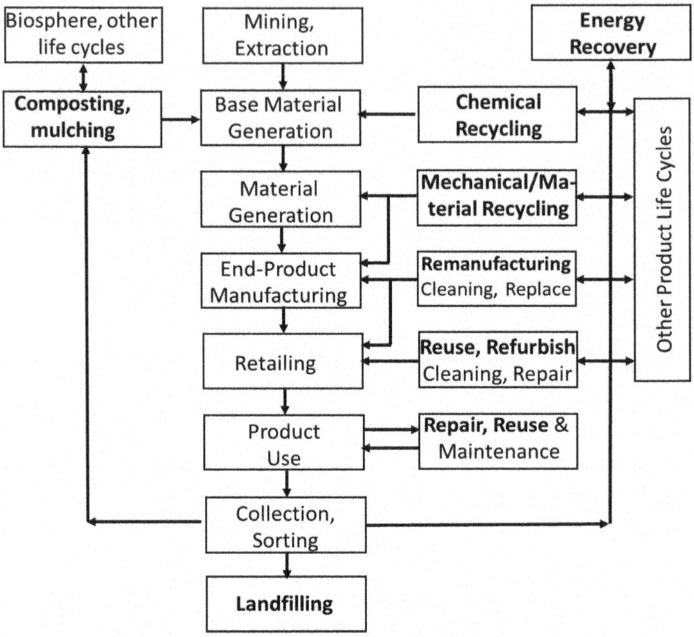

Fig. 8.1 Resource flows & Circular Economy. ©Wulf-Peter Schmidt 2024. All Rights Reserved

approaches within a product life cycle. However, it is important to note that the materials can also flow between different product life cycles allowing efficient material utilization cascades—between industries.

This chapter addresses the fundamentals of engineering before looking at different technological options in a circular economy.

8.1 Engineering Methods

8.1.1 Waste Preprocessing

Several classical engineering options for end-of-life operations exist. These prepare the waste streams for the subsequent reuse, remanufacturing, recycling, or recovery processes. The aim is to isolate recoverable components from each other to allow their easier recovery. Varying characteristics and behaviors of waste materials are used for separation of materials. In particular, distinguishable differences in form, color, density, reflecting behavior, surface properties, magnetic properties, electric properties, crushing behavior, etc. are used.

The following engineering techniques attempt to either ensure that these distinct properties and behaviors are accessible or directly exploit them in order to isolate the different materials/components:

- Cleaning
- Surface enlargement/Comminution
- Surface reduction/Agglomeration
- Sizing/Classification
- Sorting

The key performance indicator (KPI) of waste preprocessing is the separation share, S, of the target fraction, T (%). This S_T is derived from the content/share of the target fraction CT (e.g. the content of valuable recycling material or harmful substances/materials that need to be removed) in the mass m of the output (O) compared to the input material (I).

$$S_T = \frac{CT_O \times m_O}{CT_I \times m_I} \times 100 \tag{8.1}$$

Example: the separation share for the target substance copper is as follows if the copper share of 1% in 100 kg input material is increased to 50% in 5 kg output material:

$$S_T = \frac{0.5 \times 5\,\text{kg}}{0.01 \times 100\,\text{kg}} \times 100 = 250\% \tag{8.2}$$

On a material level—not considering the exact content share of valuable substances—this is simplified to a Mass separation factor

$$S_M = \frac{m_O}{m_I} \times 100 \tag{8.3}$$

This leads in the above example to

$$S_M = \frac{5}{100} \times 100 = 5\% \tag{8.4}$$

The example shows that the KPI S_T is most relevant if the target substance is significantly accumulated by the process.

Waste processing can be based on cleaning, surface enlargement or reduction, sizing, classification, and sorting. The key performance indicator is the share of the targeted material to the input.

Cleaning is usually done mechanically or chemically and rarely by other means like electrochemistry. It is a process that can use significant amounts of water or chemicals and must support the following processes while minimizing negative implications such as the need for drying processes, etc.

8.1.1.1 Comminution

Surface enlargement by comminution can be achieved by various principles:

- Mechanically stressing the materials by means of compressive forces (F_C), shear forces (F_S), impact stress (F_I), shear flows, friction forces (F_F), or cutting. Cutting with a wedge-formed cutter typically exercises a compression force (F_C) first while further penetration by the cutter exerts strong shear forces (F_S) and finally tractive forces (F_T) that split the material.

  All these different forces address the (different) resistances of the targeted materials. A fracture occurs if the local stress is greater than the molecular tensile strength. During comminution, particles of the input material are broken into many smaller particles. The breakage occurs precisely when a critical elastic stress field is created within the particles (Fig. 8.2).
- Mechanical comminution methods can be modified by adding further stresses. For example, chemicals added to the mixture before (wet or dry) grinding causes additional chemical stress on the surface of materials and eases the grinding process, improves size distribution, and reduces agglomeration. Similar principles work if the added agents are microorganisms, thermal stresses, etc.
- Thermal stress applies heat energy for example through conduction, convection, or radiation, leading to cracks formation due to thermal expansion and contraction. Thermal energy can be applied by conventional furnaces or by electromagnetic (microwave or radio wave/radiofrequency) or dielectric heating. In contrast to conventional furnaces, electromagnetic heating is not based on convection, conduction or radiation but on energy conversion: material in the electromagnetic field interacts with electric and magnetic polarization depending on the frequency of the electromagnetic device used [2].

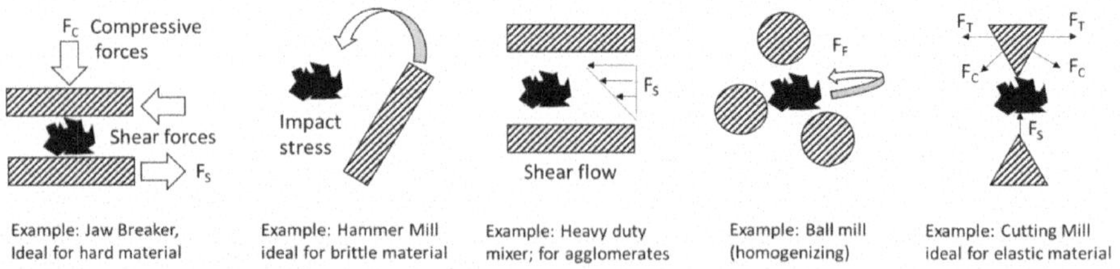

Fig. 8.2 Types of comminutions based on mechanical forces. ©Wulf-Peter Schmidt, all rights reserved

- Chemical stress can be applied by substances penetrating the material and creating fractures through internal physical bonding forces being broken up. Differences in swelling characteristics, expansions and contractions, partial dissolution, capillary stresses, and/or temperature stresses occur during the absorption of the comminution agent [3].
- Pulse stress generated by electrical, magnetic, or ultrasonic/acoustic pulses can lead to microcracks that cause comminution. Electrical stresses by high-voltage electrical pulses cause fragmentation. This approach utilizes the (different) electrical conductivities leading to expansion when a high voltage is conducted through the material. In some cases this may even lead to an electrical disintegration if the electrical treatment causes such high pressure that the tensile strength of the material is exceeded.
  Magnetic stress generated by magnetic pulses in a magnetic field interacts with magnetic waste materials. Again this can be used to create microcracks resulting from frequent changes in the magnetic dipoles of magnetic materials.
  Similarly, if three-dimensional acoustic waves above 20 kHz are passed through waste material, the wave velocity varies depending on the material, particle size, etc. causing reflection and refractions at the boundaries. This leads to tensile stresses [2].
- Deformation depends on the behavior of materials and substances. Elastic behavior corresponds to linear, reversible, time-independent behavior, where the same tensile stress-strain curve exists during loading and unloading. Ductile behavior leads to irreversible, permanent/time-independent deformation (e.g. metals). Viscous behavior leads to irreversible, but time-dependent behavior in which short-term deformation corresponds to higher tensile stress forces (e.g. an impact mill has a short loading time) (Fig. 8.3).

Ideally, selective comminution is carried out in which, for example, brittle materials are addressed by impact stress, while elastic materials are comminuted by cutting. Comminution is often done sequentially with bigger crushers/breakers first, followed by mills to get to the final particle size.

A key performance indicator of comminution is the Reduction Ratio which is the ratio between the average particle diameter before and after comminution. The distribution of the particle diameters is another important parameter that assesses quality (usually the narrower the distribution the better).

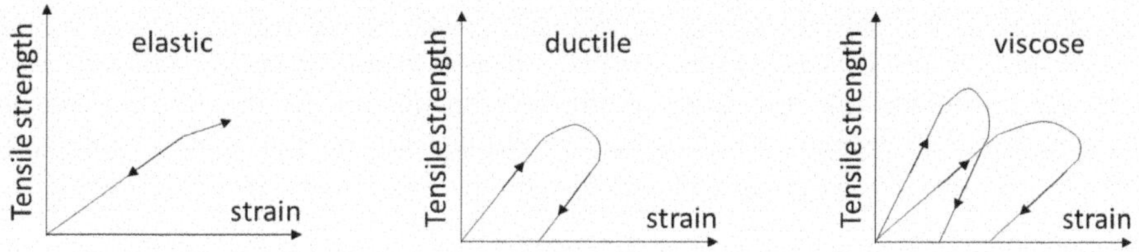

Fig. 8.3 Material behavior under tensile strength. ©Wulf-Peter Schmidt, all rights reserved

Mechanical/thermal/chemical stress, magnetic/electric/acoustic pulses, deformation lead to comminution, measured as the reduction ratio between average particle diameter before/after comminution.

Comminution is typically an energy-intensive process linked to high emissions if not sufficiently captured.

8.1.1.2 Agglomeration

Surface reduction through agglomeration can improve the handling of the waste stream, achieve lower dust emissions, and decrease the particle size distribution within the waste stream. Different methods can be used here [4]:

- Built-up agglomeration or built-up pelletizing is achieved by rotational rolling and spraying with binder fluid or steam. The rotational motion of the mixing chamber pushes the particles to the edge so that all particles move along the same path. Over time, this creates a snowball effect where particles gradually combine and diffuse to form a narrower size range.
  Granulation, pelletization, micro-pelletization, and conditioning are examples of common build-up granulation techniques. Each process differs in scope, shape, particle size distribution, and uniformity of the pellets produced. Flow mixers or ring-layer mixing granulators can be used as wet granulators or pelletizers. If particularly large agglomerates are desired, multi-stage agglomeration is carried out using a pelletizing plate.
- Pressure agglomeration. Powder or other dry particles are mechanically pressed into the solid form that is desired. The resulting force can vary in intensity: from low-pressure extrusion via high-pressure rolling to punch compaction. Pressure agglomeration is usually a dry process. Moisture or binders are not required or used in a minimal amount. Examples include roller presses and stamping presses.
- Thermal agglomeration, e.g. by sintering or fritting. Surface melting allows the formation of material bridges. These bridges are generated by atoms that diffuse within materials. Sintering is often used for materials with high melting points like metals and ceramics because materials do not need to be molten in this process.
- Spray and fluidized bed agglomeration is a process where powders are dispersed and dissolved in a liquid in such a way that they can be sprayed through a nozzle. The spraying process takes place above a heated drying or spraying tower. The sprayed droplets sink downwards in the hot, swirling air. By the time they reach the bottom, the droplets are dried out. The heated powder particles can agglomerate with other powders in a pneumatic fluidized bed as desired. Agglomerates produced in this way are usually readily soluble when mixed with liquid.

The key performance indicators for comminution are applicable for agglomeration as well.

Agglomeration is achieved by rotational motion, pressure, thermal treatment, spraying, and fluidization. It supports the handling of material streams, narrows the particle size distribution, and lowers emissions.

8.1.1.3 Classifying/Sizing and Sorting

Classifying or sizing is the separation of particles by particle size or more precisely, by the "minimum cross-section presented during the time of passage across the meshes of the screen cloth" [5]. "Classifiers are machines that sort materials according to their size, shape, and density. They can be divided into two different categories based on the technology they use. Air classifiers separate materials by utilizing a dry process. Hydro cyclones, on the other hand, sort particles in a liquid suspension" [6].

Sorting is the separation by characteristics and behavior of materials in the fluid. This combines classifying and sorting. For example, if homogenous material is run through a Hydro cyclone, the material is sized according to the particle size. However, if a mixture of materials is fed into a Hydro cyclone, materials will be sorted by their density.

Key performance indicators include the selectivity function S(c) (Eq. 8.5) which is the ratio between the density function f of the separation characteristic c of the output (O) versus the input (I) multiplied by the Mass Separation factor (see Eq. 8.3). The steeper the selectivity function in the middle of the distribution the better the sorting quality (ideally a step function).

$$S(c) = S_M \times \frac{f(c)_O}{f(c)_I} \tag{8.5}$$

Classifying separates particles by their effective sizes while sorting is based on differences in material characteristics/behavior. The selectivity function of the output to input characteristic should be steep.

A huge multitude of technologies classify and sort materials including:

- Sieves and filters of different sizes. The yield can be impacted by increasing frequency, pressure, additives that reduce adhesion between particles like hydrophobic lime, etc.

- Aero- and Hydro Cyclones separate particles by (sink) velocity in the air or liquid flow—velocity depends on particle size and density.
- Manual or automated picking requires the easy differentiation and isolation of materials/parts to separate them from the rest of the mixture. Automation can be based on sensors detecting color, reflection, dielectrical constants, electrical resistance, radiation absorption, etc. The identified parts are automatically removed by a strong air blow via a targeted nozzle directed at the material stream.
- Density sorting can be done in centrifugal fields (e.g. Hydro cyclone), counter or cross current flows, pulsating flows, in fluidized beds, in dry conditions or by a fluid with a defined density that enables density separation at heavier densities (swim sink separation). The differences in density, surface condition, particle sizes, and forms as well as numerous process conditions impact the selectivity of the sorting.
- Magnetic sorting exploits differences in magnetic properties and is especially suitable for the separation of magnetic metals as well as the separation of metals from other materials if materials are sufficiently isolated and not fixed to each other.
- Electric sorting is based, for example, on an eddy current separator. A rotating magnetic pole wheel covered by the conveyor belt generates an eddy current in conductive materials (especially metals) when the conductive materials are in an alternating magnetic field. These eddy currents in turn build up an opposing magnetic field, which leads to separating repulsions.
- Thermal sorting uses differences in melting and boiling points. This method can be very energy intensive.
- Chemical sorting exploits different chemical and physical properties of particles. For example, a particular solvent might dissolve either the targeted or the unwanted substances to allow for later separation.

Separation of mixtures exploits differences in shapes, density, magnetic/electric/chemical/thermal, and other properties. Sorting requires different parts to be sufficiently isolated and not glued together.

8.1.2 Reduce, Reuse, Recycling, and Recovery Technologies

The preprocessed waste can be the input for various recycling and recovery technologies that can be categorized into:

- Reduce and Avoidance Technologies
- Reuse, Refurbish and Remanufacturing
- Mechanical Recycling

- Metallurgical Recycling
- Chemical Recycling
- Other Recycling technologies
- Energy Recovery

The main technologies that reduce and avoid waste are already implemented as part of the product and production design phase (see Chap. 4). In addition, smart use can avoid waste.

Reuse, refurbishing, and remanufacturing are mainly based on common repair technologies as well as replacing or up-dating worn or outdated parts.

Mechanical recycling uses mechanical processes like those mentioned in Sect. 8.1.1 combined with classical production technologies like primary forming, forming, joining, coating, changing material properties, etc. (see Chap. 6). Mechanical recycling occurs on a material level with key performance indicators including the difference in quality between the recycled product(s) and the original virgin materials. In addition, the recycling rate is a key performance indicator that is regulated for some products (e.g. Batteries in Europe).

Metallurgical recycling of waste metal requires the metal's extraction from the waste input which uses:

- Thermal energy in pyrometallurgical recycling. The main environmental impact is the high energy demand which can partly be mitigated by energy recovery and catalysts that address emissions.
- Liquid/aqueous solutions and chemicals in hydrometallurgical recycling. The negative environmental impacts can be mitigated by a "circular hydrometallurgy" which closes water loops, regenerates chemicals, decreases activation energy, etc. [7].
- Reduction of metal ions to dissolved solids (or) liquids by electrochemical processes during electrolysis or by using reducing elements in electrochemical recycling.

Chemical recycling of plastics does not recycle products on a material level (plastic to plastic) as is the case in mechanical or metallurgical recycling. Instead, products are recycled on a monomer basis to form pre-products for plastics, fuels, or other chemicals. The advantage of this method is that the monomers usually have the same quality as virgin monomers while mechanical plastic recycling often requires numerous processes to obtain similar quality levels to virgin plastic (if at all possible). Various technologies exist for chemical recycling. They use chemical reactions and/or thermal energy to break-down plastic polymers into monomers. Examples include gasification, pyrolysis, depolymerization e.g. by solvent-based processes, hydrothermal microwaves, and many more.

Other recycling technologies include combinations of all of the above together with new innovations. For example, biotechnology-enabled polymer recycling uses enzymes that break-down polyester, nylon, etc. This technology breaks down polymers back into oligomers and then monomers with sufficiently high qualities to allow for re-polymerization.

Mechanical/metallurgical recycling creates recycled materials. Chemical recycling forms monomers that can be used as the basis for new plastics, fuels, or other chemicals. Quality of products is key.

Waste energy recovery only uses the heat value of waste streams leaving limited opportunities for circular economy. Nevertheless, it might be an option if the effort for further preprocessing for recycling is linked to excessively high energy demand and the resulting product could not substitute virgin materials for example for quality reasons. Energy recovery processes convert materials by the following routes [8]:

- Thermal conversion and oxidation of waste at high temperatures (incineration) to steam, purified flue gas, and ash. Important is effective emissions control even if this reduces the energy yield.
- Biochemical conversion in the form of enzymatic decomposition of organic, biodegradable waste by biological processes leading to the capture of methane gas, alcohol, and other products.
- Thermochemical conversion by pyrolysis or gasification leading, at high temperatures, to the decomposition of organic, non-biodegradable, low-moisture waste to heat or fuel oil and gas.
- Electrochemical conversion exploiting the "oxidation-reduction machinery" of immobilized microbial cells to generate electricity and hydrogen (microbial fuel cell).

The key performance indicator of waste energy recovery is the energy yield (output versus input) and the net emission throughout the life cycle considering the substituted energy carrier. From a circular economy perspective additional consideration are important (see Sect. 8.2).

Energy recovery based on thermal, bio-, thermo-, or electrochemical conversion is an option where the generation of recycled materials would not lead to sufficient levels of quality.

8.2 Sustainable End-of-Life Operations and Circularity

In general, the sustainability of the end-of-life operation depends on

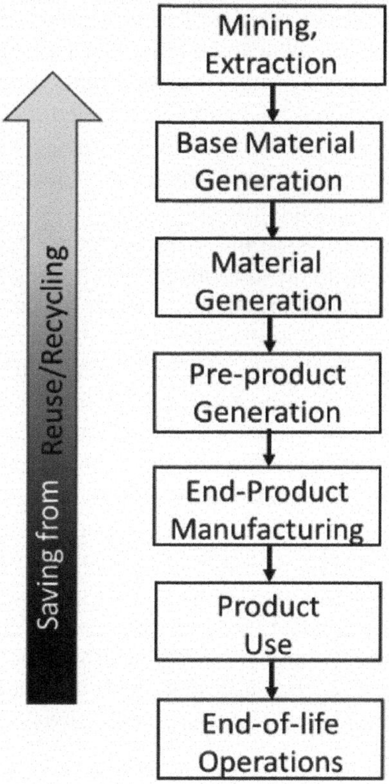

Fig. 8.4 Savings/benefits from recycling processes depending on how far back loop is closed in the up-stream life cycle

- At which stage of the product's life cycle materials are substituted for by the recycled material—this determines the impact magnitude of the substitution (Fig. 8.4).
- The quality of the generated material/product and the impacts of the substituted product (effectively the recycling/recovery credit)—this is mainly driven by the type of material, conditions of the waste material/end-of-life operation input including impurities, design, compatibility of linked materials as well as the end-of-life technology chosen and the sustainability of the substituted processes.
- The impacts of the end-of-life operations themselves and whether and to what extent these impacts and the efficiency of recycling operations are increased or decreased either by how far back the circular loop is closed or by how high the recycling/reuse rate is.

The further back the circular loop goes in the value chain the lower the potential savings from circularity. This rule is modified by the quality of the recyclates & the impacts of the end-of-life operations.

A perfect, 100% circular economy has practical limits. This is partially due to the technical and quality constraints mentioned above, but also because some materials are combusted (e.g. fuels), or are lost by other means (corrosion, dissipation etc.), making them unavailable for circularity. More importantly, circular economies might not be socially beneficial if local circularity targets have adverse impacts beyond borders, which can include exacerbating global power imbalances in lower-income countries, such as exploitative labor practices [9]. There are examples for both cases: electronic waste centers in Africa increase employment and reduce emissions from toxic substances by recycling in controlled environments while irresponsible recycling practices endanger the health of workers and surrounding communities.

There is an optimum recycling/reuse/recovery rate from a sustainability perspective [10]. This is mainly because the effort of recycling usually increases the higher the recycling rate is. This trend exists as higher effort is required to ensure the secondary material has adequate quality levels as well as due to the logistical challenges presented by collecting the waste materials. Paper or other organic material recycling is a common example for the increasing effort in recycling processes with higher recycling rates. The higher the recycling rate, the higher the sorting, filtering, and cleaning efforts. For paper, the intensity of de-inking processes and the amount of de-inking chemicals increase because ink accumulates when paper cycle is closed more [11]. Figure 8.5 demonstrates the principle of distribution factors between input and output materials changing depending on the recycling rate.

Due to a decline in quality, the specific environmental benefit of additional recycling (increased recycling rates) can decrease as well. That may happen in case of:

- A substitution factor lower than 1. The substitution factor indicates the ratio of the substituted mass of primary material to the mass of secondary material in the

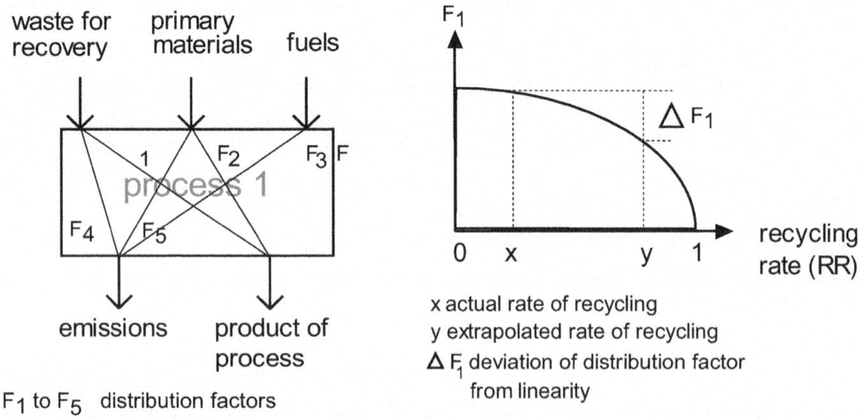

Fig. 8.5 Distribution factors between in- and output (left) and dependency of the distribution factors on recycling rate [12]

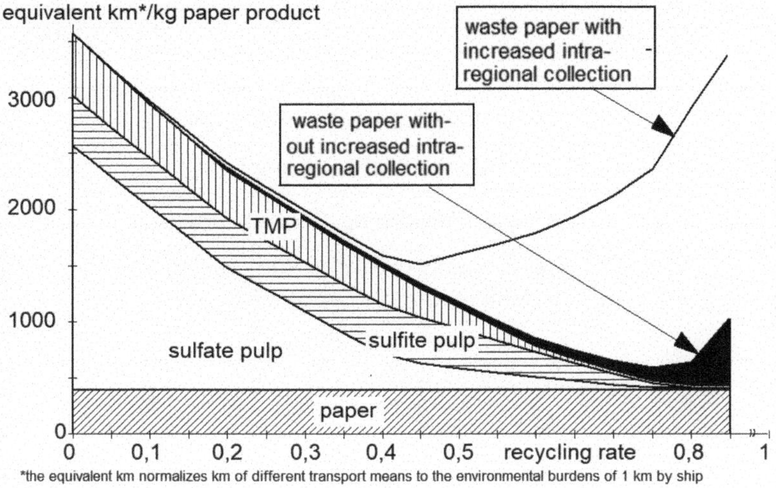

Fig. 8.6 Necessary long-distance transport in the paper recycling system for main materials (without and with increased intra-regional collection of waste paper) [12]

recycling product, assuming equal functional properties of the end product (see Eq. 8.6). An example is the reduced strength of secondary reinforced plastics that can only be compensated by thicker products, requiring more (secondary) plastics.

$$\text{Substitution factor} = \frac{\text{mass of substituted primary materials}}{\text{mass of secondary materials}} \tag{8.6}$$

- Quality conservation means. Additional processes (e.g. thermal treatment for foundry sand recycling) or more inputs and outputs in the recycling system may occur to ensure the same function as the primary product (see above example of paper recycling and Fig. 8.5).
- Reduced substitution potential because of reduced quality. The waste product can only be recycled to other products that may have lower substituted environmental impacts than products of higher quality grade.
- Expansion to mixed fractions, highly polluted fractions, and composites to be recycled.

Similarly, the impact of recycling on logistics depends on the individual case and the recycling rate. The general assumption is that recycling increases independent from imports which suggests material flows occur more within a country or region leading to lower overall transportation efforts. However, this is not necessarily always the case from a holistic life cycle perspective (Fig. 8.6).

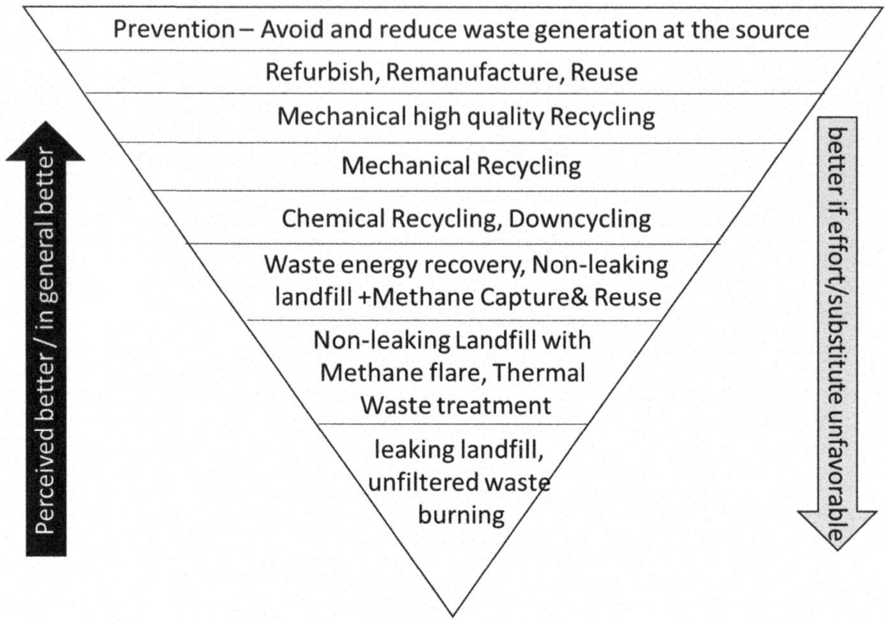

Fig. 8.7 Extended waste hierarchy—not always black-and-white. ©Wulf-Peter Schmidt, all rights reserved

There is an optimal rate of recycling which depends on the impacts of recycling/recovery processes that may change with increasing recycling rates as well as the properties of the substituted product.

These holistic, life cycle considerations also explain deviations from the generally assumed rule of a waste hierarchy [13] which suggests that reducing/avoiding waste is being better than recycling waste, and recycling waste is better than disposal or thermal energy recovery (Fig. 8.7).

8.3 Recycling of Natural Materials

Harvesting waste can be used to keep carbon out of the atmosphere (Sect. 6.9) but it can also deliver other benefits for the environment (see also Fig. 2.7). However, if such waste is just burned on fields, recycling of these residues might be the more sustainable option.

For example, more than 7 million tons of pruning residues from olive trees are generated in Europe of which approximately 75% are used to mulch the soil and the rest is burned without any energy recovery. In an EU-funded project, a process has been developed that generates a blend material made of olive tree fibers and a recycled Polypropylene (PP) matrix to substitute oil-based raw materials [14, 15]. The pruning residues are collected and grinded. The material is sieved, pretreated (with NaOH, AcOH) for hydrolysis (in the pilot stage at 90 °C for 90 min with relatively high energy and water consumption), treated in hydrolysis and finally filtrated.

The resulting material is mixed with recycled propylene and processed with twin screw extruders to create a recycled material that has mechanical properties allowing it to replace virgin, filled PP materials in injection molded automotive interior parts like footrests and panels (although for some parts the brittleness is too high for crash safety requirements), allowing a share of natural materials up to 40%.

This example of using renewable and recycled materials demonstrates that opportunities to replace fossil-energy-based parts also exist, allowing a move towards a carbon neutral product. However, these materials also cause carbon emissions if not produced with renewable energy sources alone. Therefore, the sustainability of recycling natural materials has to be checked in each case and compared to the use of conventional materials. If there are relatively low negative impacts of the substituted material, it is important that the impacts of recycling processes remain significantly smaller.

In the example of recycling of pruning residues, the life cycle comparison of the current situation with the recycling method must consider systems with comparable functionalities and benefits (compare Fig. 2.5). If only those pruning residues are recycled that would otherwise be burnt on the fields, the system boundaries for the LCA comparison are as shown in Fig. 8.8 excluding the gray parts. The environmental benefit of the recycling system would decrease (and potentially become negative) if all pruning residues were used differently. In that case, the recycling system would require an alternative production of mulch. A similar or larger deterioration of environmental benefits of recycling would occur if the resulting recycling product had a worse performance than the virgin product—especially if it adds weight. This is not the case in this example as the relevant properties of the recycled and virgin products are essentially identical.

> Treatment options for natural material residues such as recycling, carbon removal, compost generation etc. need to be analyzed for their sustainability benefits in each case.

8.4 Recycling of Fiber Reinforced Polymers

Light materials like fiber-reinforced polymers have positive environmental and economic impacts if they are used in moving parts (for example in the mobility industry) but also in other sectors like wind power generation. The third biggest

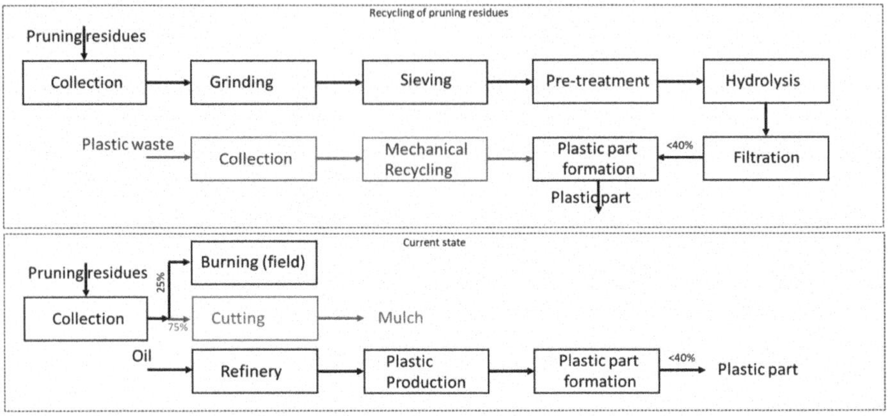

Fig. 8.8 Recycling of pruning residues from olive trees versus current state. Wulf-Peter Schmidt 2024. All Rights Reserved

application for glass fiber-reinforced polymers is Printed Wired Boards (PWBs). Most fiber-reinforced polymers end up in landfills, globally. Where mixed waste recycling exists, fiber-reinforced polymers are usually treated by mechanical grinding and sink float separation [16]. However, due to their worse material properties, the products of mechanical recycling of fiber-reinforced polymers compete with lower-cost virgin filler alternatives like calcium carbonate or silica [17]. Direct energy recovery is feasible only for aramid- or natural fiber-reinforced polymers. This is not possible for glass, basalt, or carbon fiber reinforcements for which mechanical pre-conditionings must still be developed. These fibers will end up in the slag of normal incinerators and/or disturb the technical processes. A better recycling of glass or carbon fibers can be achieved using the following approaches [16–18] (Fig. 8.9):

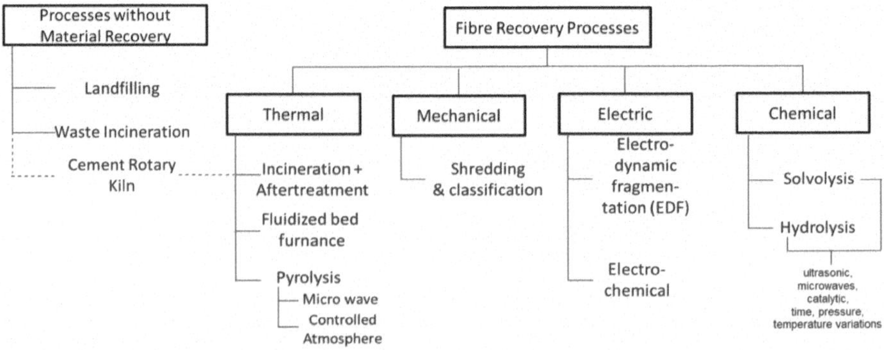

Fig. 8.9 End-of-Life options for fiber-reinforced plastics. ©Wulf-Peter Schmidt, all rights reserved

- Mechanical recycling with different shredding and grinding steps, combined occasionally with classification via cyclons and sieves. Nevertheless, this leads to 50 to 60% worse mechanical properties than virgin fibers.
- Electrical recycling (high-voltage pulses of 50–200 kV dissolve fibers from matrix) or electrochemical recycling (low voltage with carbon fiber waste as anodes, stainless steel as cathodes, and Sodium Chloride as electrolyte) results in adequate material properties of recycled fibers.
- Thermal recycling based on 3 separate methods. Worsening properties can be partially compensated by chemical treatments (e.g. chemical etching/ salinization):
  - Combustion incl. energy recovery at 450–700 °C for carbon fibers or 300–600 °C for glass fibers.
  - Fluidized beds at 450–550 °C where the matrix is burned.
  - Pyrolysis without oxygen at 400–700 °C where the matrix polymer is turned into oil and gas which is then used for the process under controlled atmosphere with vacuum, nitrogen, argon or super-heated steam.
- Chemical (solvolysis) approaches using, for example, nitric acid at different temperatures or supercritical fluids. (Supercritical and subcritical) water (hydrolysis) is also used. In an LCA study [19], supercritical water (SCW) was shown to have the lowest cradle-to-gate production energy intensity (9.5 MJ/kg) and the lowest environmental footprint. Binary mixtures of pure solvents and supercritical water have better results in terms of yields and environmental criteria.

A literature review [17] suggested profitable prices of recycled carbon and glass fibers that can compete against virgin fibers only if there are small property losses. However, the review also revealed that the environmental benefits of these fiber recycling approaches are inversely proportional to the profits of the recycling industry. Other sources suggest that it is possible to reduce the costs of recycling by downcycling the recycled material or by energy recovery, even though the production of high-performance recycled materials is usually favored [18]. The literature review concluded that a "chemical recycling process using solvents such as water at critical conditions and a binary fraction of mild solvents with water at critical conditions is considered to be the future of recycling both CFRP and GFRP waste. The maximum resin elimination ratio, the higher retention of mechanical properties in recycled fibers and the use of cheap and sustainable solvents make the process distinct from any other recycling process. However, the process is not yet commercialized, and critical operating conditions tend to consume additional energy" [17].

> Fiber-Reinforced polymers can be recycled in mechanical (low quality), electric or chemical (high-quality fiber, matrix lost, not industrialized) processes; no energy recovery for inert fibers.

8.5 EV Lithium-Based Battery Recycling

8.5.1 Background

Electric Vehicle Battery production will ramp up significantly. For Europe alone, the forecast is that there will be a 900 GWh cell production capacity by 2030 and around 6000 kt of end-of-life batteries by 2040 [20]. Although there is currently a discussion about a weakening of the increase of customer demand, the trend continues to be positive. Furthermore, the EU Battery Regulation [21] requires minimum recycled content shares of 6% for Lithium and Nickel and 16% for Cobalt by 2031, which by 2036 will increase to 12% (for Lithium), 15% (for Nickel), and 26% (for Cobalt). To comply with these regulations, pre-consumer recycled content is also allowed, i.e. recycling of production waste. This requirement applies to all batteries sold in the EU single market. In the same regulation, requirements for the recycling processes themselves are also defined. On one hand, a minimum recycling efficiency is mandated—defined as the ratio between the mass of output fractions accounting for recycling and the mass of the waste batteries' input fraction (a delegated act will detail the methodology). For Lithium-based batteries this efficiency target is 65% by the end of 2025 and 70% by the end of 2030. On the other hand, material recovery targets are set for several materials (Table 8.1). China and South Korea also have clear targets for the recycling of batteries including recycling rates of around 90% while the United States of America still do not have general requirements in place for battery recycling.

Given the durability of vehicles and their batteries, there is a time gap of 12 to 15 years between the ramping up of electric vehicles and the increase in end-of-life batteries on the market. This is the reason for low recycling capacities in the early 2020s (with recyclable input materials being only around 10 kt by 2023, of which 70% is production scrap) and the focus on production waste. Increases in capacity will occur during the 2030s (recyclable materials of around 460 kt by 2030 of which 60% will be production scrap and 2650 kt by 2035 with 15% production scrap) [20].

Table 8.1 Material recovery targets of EU Battery Regulation

	Compliance by end of 2027	Compliance by end of 2031
Cobalt	90%	95%
Copper	90%	95%
Lithium	50%	80%
Nickel	90%	95%

Regulations mandate recycling of EV batteries. The 12+ year time gap between production and end-of-life recycling leads to delayed capacity build-ups and low availability of recycled content.

8.5.2 Recycling Technology

There are different recycling options for EV batteries that start with different types of preparation and pre-treatment. The first step is usually to discharge the battery for safe handling, although experts (e.g. Hydromet) suggest this might not be needed for some recycling processes e.g. in wet inert nitrogen atmospheres. In most cases, the battery must be dismantled by removing it from the end-of-life vehicle. The exception to this is the recycling of battery production scrap. The battery housing, frame, wiring, and cooling system is removed in this step, separating the individual battery modules and cells for further processing while the removed iron, aluminum, copper, and plastic parts can be recycled separately. There might be the thermal treatment of the modules and cells to prepare for comminution/crushing of cells by various methods like shredding (dry or wet), impact mills, or shockwaves. The final pre-treatment step is the generation of the so-called black mass by different separation processes that can include sieving, froth flotation, density, and magnetic separation.

This black mass can be sent to central recycling processes that require higher input volumes to be efficient. Currently, there are different recycling options [20, 22, 23]:

- Pyrometallurgy (smelting) is the most developed (and most capital intensive) recycling route. It separates metals by melting point and thus requires relatively high temperatures of around 1400–1700 °C. Therefore, there is a high energy consumption to separate cobalt, nickel, and copper. Reducing agents like activated carbon, aluminum foil, $NaHSO_4$, $CaCl_2$, and/or NH_4Cl are also used. However, Lithium and Manganese end up in the slag together with slag modifiers like Al_2O_3, CaO, MgO and/or SiO_2. To recover these materials, further processing is necessary.
- Hydrometallurgical recycling involves the black mass undergoing leaching and extraction processes. First, Graphite is leached out and can be refined to battery grade material quality in a post-treatment step. A second leaching step removes impurities, leaving behind Manganese Sulfate, Cobalt Sulfate, Nickel Sulfate, and Lithium Carbonate. Leaching agents may include Lactic acids, H_2O_2 etc. Companies like Hydromet claim to recover more than 95% of these four materials (Manganese Sulfate, Cobalt Sulfate, Nickel Sulfate, and Lithium Carbonate). The best results can be achieved if the exact battery chemistry is known which is

supported by the EU Battery regulation that requires individual Battery Passports. However, this requirement may lead to recycling efficiency limits when scaling-up the process.

- Direct recycling is the least mature technology option but has the potential to recover cobalt, lithium, nickel, manganese, aluminum, and copper from the cathode as well as graphite from the anode side with lower environmental impacts. Like hydrometallurgical recycling, the best results can be achieved if the specific battery chemistry is known. Direct recycling can be based on a combination of thermal and chemical processes that do not disturb the original compound's structures or break the materials down into individual elements. The separated parts with the purified active materials may show surface and bulk defects that are repaired by relithiation (regeneration by adding new lithium ions) or hydrothermal processes (4 h at 220 °C, others work at 80 °C for 6 h) followed by an annealing (800 °C, 4 h) or ultrasonic power (600 W). Some processes regarding the purification of the graphite anode to enable its recovery have been proposed. Research aimed at recovering other battery parts (electrolyte, etc.) is also ongoing. The advantage of this route is that cathode (and potentially electrolytes and anode) materials can be used again directly (Fig. 8.10).

In summary, the maturity of processes is currently inversely proportional to the recycling efficiency in terms of the materials that can be recovered and the quality of recovered materials.

Pyrometallurgical battery recycling is the most mature option but cannot recover Lithium / Manganese. Hydrometallurgical & direct recycling recover all materials at a high quality.

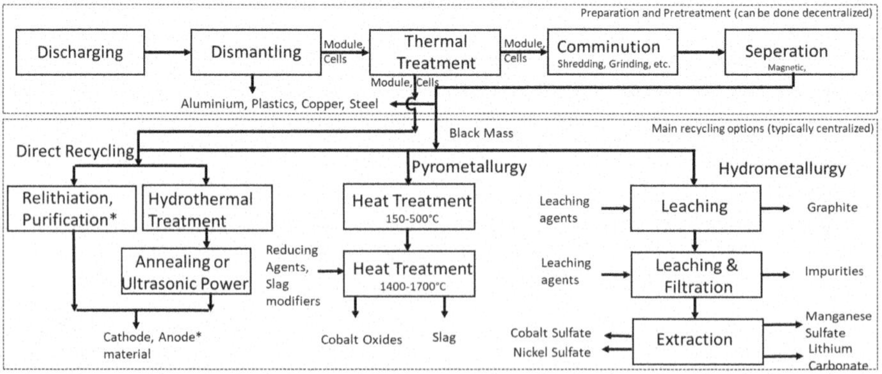

Fig. 8.10 EV Battery Recycling options (simplified overview). ©Wulf-Peter Schmidt, all rights reserved

For practical implementation, the logistics of end-of-life (EoL) EV batteries is also important. In a study considering the economy of scales and optimization of transportation in hydrometallurgical recycling, it was recommended to have local recycling capacities of 10,000 tons per year to pre-treat EoL batteries to Black mass (getting around 4000 tons per year) and transport from around 10 local facilities to one recycling hub with capacities of around 40,000 tons per year to generate around 19,000 tons of battery grade materials [20]. Considering the announced investments of 2.2 bn Euros, the study identified an existing excess capacity especially for pre-treatment spokes (but also for recycling hubs) up to 2030. Afterward, additional investments of 7 bn Euros will be needed, however, technological changes may change the anticipated costs. The initial excess capacity reflects the low number of EV end-of-life vehicles: the potential recycling input material remains limited up to 2030. The study calculated that sufficient recycled material will be available to support the recycled content mandates apart from a shortage of recycled cobalt in the early 2030s.

8.6 Recycling of Rare Earth Containing Magnets

The EU Joint Research Center (JRC) determines a lack of circularity for materials and components, explicitly mentioning rare earth permanent magnets (REPMs) that are basically lost in the shredding process. The JRC estimates the potential for collected and disassembled REPM e-motors to be around 0.7 kilotons (kt) of REPMs in 2030 and 4.2 kt in 2040 [24]. The JRC estimates that targeted dismantling and recycling or reuse could improve greenhouse gas emissions only slightly (4–15%) and have limited socio-economic benefits while leading to added costs and positive impacts on resource scarcity (mainly due to the recycling of copper from e-motors). However, if combined with design for disassembly the JRC estimates that targeted dismantling and recycling or reuse could improve greenhouse gas emissions only slightly (4–15%) and have limited socio-economic benefits while leading to added costs and positive impacts on resource scarcity (mainly due to the recycling of copper from e-motors). However, if combined with design for disassembly (see Sect. 4.1.1), the environmental and socio-economic benefits are greater. There are challenges in removing the magnets from e-motors (due to accessibility, design restrictions, and the magnets automatically fixing themselves to the surrounding metals during dismantling).

Once the magnets are separated, there are different methods to recycle rare earth from magnets [25]:

- Direct reuse. This is typically limited to large, easily accessible magnets. An example is the global recycled magnet program of 5 global automotive manufacturers in which motor rotors are dismantled, the aluminum cover separated, and a thermal treatment (above 400 °C) demagnetizes and degums the material to allow for easier handling. The magnets are then separated and cleaned, creating clean material that can be magnetized again.

However, the dimensions of the old magnets might not be appropriate anymore. In addition, a simple reuse might be a sub-optimal use of heavy, expensive Rare Earth Metals because of more efficient uses of these metals today (legacy magnets have a content of 5–6% of heavy rare earth metals while today's magnets often have only between 1 and 2%).

Direct reuse of magnets might be an option for large magnets but does not leverage efficiencies in the use of less heavy rare earth shares per magnet—different to material recycling.

- Magnet to Rare Oxide recycling by hydrometallurgical recycling (Fig. 8.11) or chromatographical recycling. The advantage of the hydrometallurgical process is its applicability in all magnet compositions including oxidized magnets.
  In hydrometallurgical recycling, rejected magnets and end-of-life magnet materials are first grinded using a ball grinding process. The grinded material is then calcinated in a rotary kiln to remove impurities at high temperature. The output is further grinded by a Raymond mill to below 0.05 mm sized particles that can be dissolved with hydrochloric acid, ammonia water, and catalyst. This step will separate the resulting iron slag and other solid precipitate from the rare earth metal containing solution. This rare earth metal solution is mixed with an extraction liquid (e.g. P507), hydrochloric acid, and ammonia and moves through several tanks to separate the Rare Earth elements in chloride form by atomic number. During this step, the purity of heavy earth elements already exceeds 99.9%. The subsequent precipitation process adds Ammonium Bicarbonate (NH_4HCO_3) and Oxalic Acid ($H_2C_2O_4$) to the solution. After a period of standing, rare earth carbonate is produced: Ammonium Bicarbonate produces the rare earth carbonates $PrNd_2CO_3$, Ce_2CO_3, and Gd_2CO_3 while Oxalic Acid leads to $Tb_2(CO_3)_3$ and $Dy_2(CO_3)_3$ formation. These Carbonates or Oxalates are transformed to rare earth metal oxides by calcination at around 1000 °C during which the CO_2 is liberated and rare earth element oxides with purities similar or even better than virgin sources are left behind. However, hydrometallurgical recycling requires many steps and high amounts of chemicals leading to major wastewater streams.
- Chromatographical recycling starts by demagnetizing (at 300 °C), crunching, and grinding magnets to create a powder form. The powder is treated with heat (around 700 °C) to oxidize the iron. Then, the heated powder is passed into a reactor where the solution is pressed through chromatography resin columns. Sulfuric gas removes rare earth elements into the leachate to obtain a 95% rare earth purity. Finally, chemicals (like recycled EDTA) wash out the rare earth from the leachate as rare earth sulfates that are turned to rare earth oxides (>99.5 purity; with 1.3 kg CO_2/kg rare earth oxides).

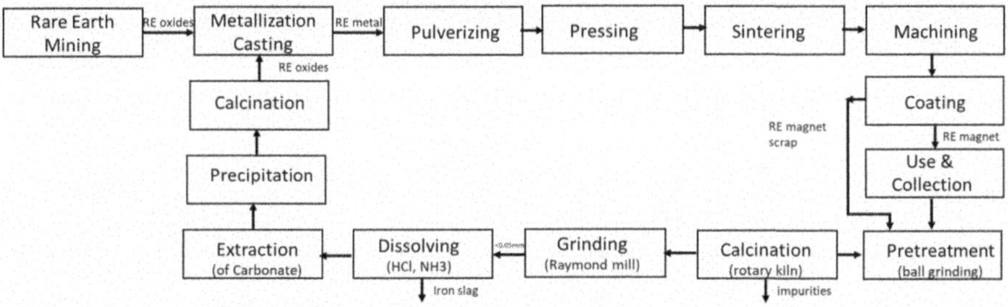

Fig. 8.11 Magnet to Rare Earth Oxide hydrometallurgical recycling. ©Wulf-Peter Schmidt, all rights reserved

> Hydrometallurgy is suitable for all magnets and leads to quality levels comparable to virgin material. Chromatography leads to slightly lower purity levels but at lower negative impacts.

- Magnet to Rare Oxide recycling by pyrometallurgical recycling has less process steps than hydrometallurgical recycling but is more energy intensive as it includes a 30 min oxidation process at 700 °C followed by a reduction process with graphite in an Argon atmosphere at 1500 °C. It leads to the formation of a rare earth oxide containing slag and metallic iron (not just iron slag like in the hydrometallurgical recycling). The slag is further refined [26]. Pyrometallurgical recycling is not suitable for oxidized magnets.
- Direct physical recycling (Fig. 8.12).
  The hydrometallurgical, chromatographical, and pyrometallurgical recycling of magnets leads to the formation of Rare Earth oxides which is not very helpful from a CO_2/energy perspective because it only substitutes the mining process (and not the energy-intensive production process). Mining itself is not the most energy-intensive step but carries eco-toxicity risks. The energy-intensive metallization process from oxide to rare earth metal is not prevented and the recycling process itself requires many energy-intensive process steps (compare Fig. 8.4). Direct physical recycling which does not produce rare earth oxides but recovers rare earth metals is also an option.
- Here, magnets are commonly glued together and can be separated by thermal treatments of up to 250 °C (demagnetization does not occur until 400°). In the traditional physical recycling process, a strip smelting process (over a water-cooled copper roll) creates thin strips that are treated with hydrogen. The recycled product is a rare earth metal that can either be directly fed back into the pulverizing process (although the share of recycled materials is limited to a maximum of 20%) or into a controlled blending process where the heavy earth metal share is diluted with virgin materials (see dotted lines in Fig. 8.12).

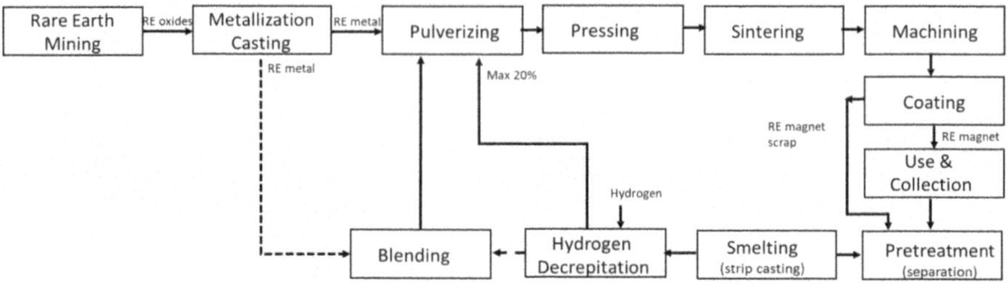

Fig. 8.12 Magnet to Rare Earther Metal physical recycling. ©Wulf-Peter Schmidt, all rights reserved

Pyrometallurgy is not suitable for oxidized magnets. Direct recycling of magnets reduces impacts most because it replaces the more energy-intensive rare earth metals and not oxides.

Greater environmental benefits exist if magnets are recycled to rare metals rather than rare earth oxides. For example, the carbon footprint of virgin PrNd is 35.19 kg CO_2/kg while it is 8.85 kg CO_2/kg in the case of recycled material (numbers for Dy are 37.97 kg CO_2/kg virgin and 11.63 kg CO_2/kg recycled or 66.8 kg CO_2/kg virgin Tb and 40.46 kg CO_2 /kg recycled Tb). Physically recycled magnets may have a lower quality if high-end magnets are recycled.

References

1. ISO 59020, *Circular Economy—Measuring and Assessing Circularity* (International Standards Organization, 2021)
2. S.O. Adewuyi, H.A.M. Ahmed, H.M.A. Ahmed, Methods of Ore Pretreatment for Comminution Energy Reduction. Minerals **10**(5), 423 (2020)
3. A. Ansems, R. Bloebaum, Chemical Comminution, in *Powder Technology*, vol. 40, (Elsevier, 1984), pp. 265–268
4. Amixon, Agglomeration [Online], https://www.amixon.com/en/glossary/agglomeration-granulation#:~:text=There%20are%20three%20main%20industrial,sintering%20and%20build%2Dup%20granulation.&text=In%20pressure%20agglomeration%2C%20powder%20or,into%20the%20desired%20solid%20form
5. E. Pryor, Industrial Sizing and Sorting, in *Mineral Processing* (Springer, Dordrecht, 1965), pp. 179–232
6. Metso, Classifiers for Dry and Wet Separation [Online], https://www.metso.com/portfolio/classifiers/.
7. K. Binnemans, P.T. Jones, The Twelve Principles of Circular Hydrometallurgy. J. Sust. Metall. **9**, 1–25 (2023)
8. EAI, Technological Options for Energy Recovery from Waste [Online], https://www.eai.in/ref/ae/wte/pro/process.html
9. Circle economy, *The Circularity Gap Report 2023* (Circle Economy, Amsterdam, 2023)
10. W.-P. Schmidt, *Ökologische Grenzkosten der Kreislaufwirtschaft* (TU Berlin, Berlin, 1997)

11. W.-P. Schmidt, G. Fleischer, High Recycling Rates = High Reduction of Environmental Burdens?, in *R´97 Congress Proceedings, 4–7 February 1997*, vol. II, pp. 137–142 (1997)
12. W.-P. Schmidt, *Ökologische Grenzkosten der Kreislaufwirtschaft* (Berlin, 1997)
13. European Union, Consolidated Text: Directive 2008/98/EC of the European Parliament and of the Council of 19 November 2008 on Waste and Repealing Certain Directives (Text with EEA Relevance) (Brussels, 2008)
14. Life Compolive—New Generation of bioCOMPosites Based on OLIVE Fibers for Industrial Applications (23 January 2024) [Online], https://www.lifecompolive.eu/
15. I. Wehmayer, T. Baranowski, COMPOLIVE—Towards Automotive Applications, in *BioSC Symposium* (Bonn, 2023)
16. Umweltbundesamt (German Federal Environmental Protection Agency) (3 February 2020) [Online], https://www.umweltbundesamt.de/faserverbundwerkstoffe-zukunftsmaterial-offener#faserverbundwerkstoffe-und-ihr-potenzial-fur-klima-und-ressourcenschonung
17. S.K.T. Karuppannan Gopalraj, A Review on the Recycling of Waste Carbon Fibre/Glass Fibre-Reinforced Composites: Fibre Recovery, Properties and Life-Cycle Analysis. SN Appl. Sci. **2(3)** (2020)
18. R.J. Tapper, M.L. Longana, A. Norton, K.D. Potter, I. Hamerton, An Evaluation of Life Cycle Assessment and Its Application to the Closed-Loop Recycling of Carbon Fibre Reinforced Polymers. Compos. Part B Eng. **184**(107665) (2020)
19. Y. Khalil, Sustainability Assessment of Solvolysis Using Supercritical Fluids Forcarbon Fiber Reinforced Polymers Waste Management. Sustain. Prod. Consumption **17**, 74–84 (2019)
20. Strategy & (PriceWaterhouseCooper), PEM (RWTH Aachen), EU Recycling Market—The EU Recycling Market—A Viable and Sustainable Business (August 2023)
21. European Union, Regulation (EU) 2023/1542 of the European Parliament and of the Coouncil of 12 July 2023 Concerning Batteries and Waste Batteries, Amending Directive 2008/98/EC and Regulation (EU) 2019/1020 and Repealing Directive 2006/66/EC. Official Journal of the European Union (28 July 2023)
22. K. K. Jena, A. AlFantazi, A.T. Mayyas, Comprehensive Review on Concept and Recycling Evolution of Lithium-Ion Batteries (LIBs). Energy Fuels **35**(22), 18257–18284 (18 November 2021)
23. ReCell, Direct Recycling of Materials [Online], https://recellcenter.org/research/direct-recycling-of-materials/#:~:text=Cathode%20Relithiation%3A%20Develop%20an%20energy,revive%20their%20high%20electrochemical%20performance
24. N. Tazi, M. Orefice, C. Marmy, Y. Baron, M. Ljunggren, P. Wäger, F. Mathieux, *Initial Analysis of Selected Measures to Improve the Circularity of Critical Raw Materials and Other Materials in Vehicles* (Publications Office of the European Union, Luxembourg, 2023)
25. B.S. JL Mag, *Sustainability—Deep Dive* (Publications Office of the European Union, 2023)
26. Y. Bian, S. Guo, K. Tang, L. Jiang, C. Lu, X. Lu, W. Ding, Recovery of Rare Earth Elements from NdFeB Magnet Scraps by Pyrometallurgical Processes, in *Rare Metal Technology 2015* (Springer, Cham, 2016), pp. 239–248

Chapter 9
Sustainable Stakeholder Relations

Abstract UN Sustainable Development Goal (SDG) 17 demands "*partnerships for the goal*" and revitalizing the global partnership for sustainable development. The conviction is that sustainability can only be achieved by global partnership. On an organizational level, this can be best achieved by establishing or repurposing stakeholder relations for sustainability—internally and externally along the life cycle.

Sustainable life cycle stakeholder relations are a prerequisite for holistic and efficient progress towards sustainability. The intention is to establish information and material flows as well as partnerships between the life cycle stakeholders. Fostering acceptance by and support from all stakeholders for sustainability initiatives or transformations (e.g. unions/employees, society/customers, investors) is crucial. Negative reactions or instances/news that could damage the reputation of the organization (e.g. in the supply chain) must be avoided. This chapter describes a systematic approach that (1) identifies relevant stakeholders, (2) establishes an enabling operating environment in the own organization and (3) implements sustainable transformation.

This chapter focusses on three areas. First, opportunities for sustainability in employee relationships are described in terms of a more sustainable work force and fostering employee engagement in sustainability. Second, investors and the finance sector as key external stakeholders beyond immediate life cycle stakeholders already covered in Chapters 5, 7, and 8 are considered. These parties are key in supporting the required transformation towards sustainability. Lastly, the importance of external advocacy in shaping the regulatory environment towards more enabling frameworks is considered.

Keywords Stakeholder analysis · Multistakeholder dialogue · Interest analysis · Power analysis · SMART objectives · Human resources · Investors · Regulation · Taxonomy · Carbon pricing · Environmental taxation · Lobbying · Advocacy

UN Sustainable Development Goal (SDG) 17 is about "*partnerships for the goal*" and revitalizing the global partnership for sustainable development. The conviction is that sustainability can only be achieved by global partnership. On an organizational level,

© The Author(s), under exclusive license to Springer Nature Switzerland AG 2024
W. Schmidt, *Solutions For Sustainability Challenges*,
https://doi.org/10.1007/978-3-031-63624-0_9

this can be achieved best by establishing stakeholder relations for sustainability—internally and externally along the life cycle.

Sustainable life cycle stakeholder relations are a prerequisite for holistic and efficient progress towards sustainability. The intention is to establish both information and material flows as well as partnerships between the life cycle stakeholders (see Table 2.13). This also includes fostering acceptance by and support from all stakeholders for sustainability initiatives or transformations (e.g. unions/employees, society/customers, investors) while avoiding any negative reactions or instances/news that could damage the reputation of the organization (e.g. in the supply chain).

Sustainable value creation is best done collectively and interorganizational. It should involve stakeholders from each of the different product's life cycles and life cycle stages [1]. Transformation or changes towards sustainability can only happen if stakeholders do the necessary things. Therefore, sustainable stakeholder relationships are key for sustainable business models of organizations. Considering interests and value propositions of various stakeholders will improve the value creation overall and therefore enhance the individual organization. Different stakeholders sometimes value different aspects (environmental improvements, economic value, political values like supporting regional or national objectives, social value in terms of job creation, image/brand values, etc. [1]). These aspects must be optimized or at least taken into account.

The starting point to establishing sustainable stakeholder relationships is a stakeholder analysis. This includes all directly and indirectly affected stakeholders, characterizes their interests and values in the sustainability transformation in terms of the opportunities/transformational potential and risks they face, their readiness for transformation, their value propositions and interests (including alignment with sustainability) as well as their roles and dynamics in the current and future transformed system. In a second step, the interactions between the stakeholders and their individual influence on others (influence mapping, power analysis [2]) is determined. Different roles in the system (regulator, decision maker, guardian, owner, advocate, catalyst or blocker, winner or loser, seller or buyer) [3] can help divide the stakeholders into their interest in sustainability and potential value in or helpfulness for the sustainable transformation. The power of stakeholders can be determined from the nature of the organization (e.g. controlling regulations, budgets, public opinion) or from informal settings (personal authority, friendships, joint experiences, etc.).

It should be noted that conducting a stakeholder analysis requires listening, respect and awareness of other stakeholders as well as cultural, gender, socio-economic, and educational backgrounds.

> A stakeholder analysis determines the stakeholder readiness, value, helpfulness, interests, and roles in terms of sustainability and its required transformation as well as their influence.

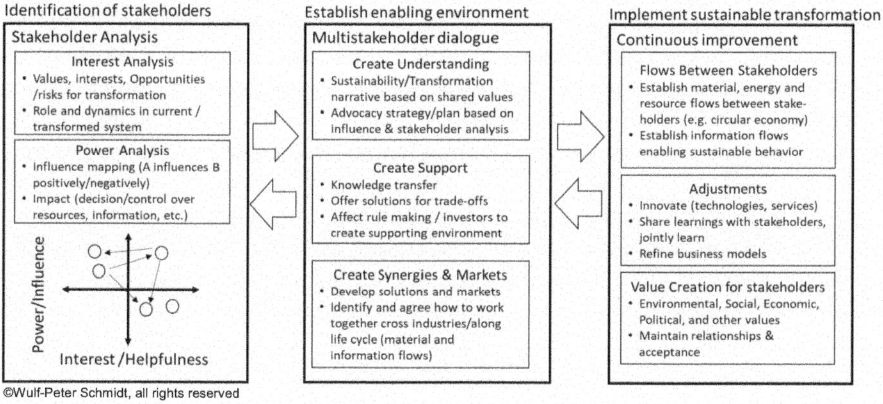

Fig. 9.1 Sustainable stakeholder relationships

Based on the stakeholder analysis, the stakeholder relationship must be established in such a way that it enables the desired sustainability transformation. This requires a multi-stakeholder dialogue. Firstly, this entails a convincing narrative/line of advocacy that garners support for the transformational vision (creates understanding and legitimacy for sustainable transformation). This narrative must start at the current level of understanding of each stakeholder and should be based on the identified values and opportunities of the stakeholders. In a multi-stakeholder dialogue, knowledge should be transferred in all directions and trade-offs should be pro-actively addressed by offering solutions for these conflicts. This will help create a supportive regulatory and societal environment. On this basis, stakeholders can establish synergies (e.g. in circular economies between industries or life cycle stakeholders). This multi-stakeholder dialogue requires a continuous engagement and initial stakeholder analysis must be updated to ensure the input from the multi-stakeholder dialogue strategy is accurate.

> Based on the insights of the stakeholder analysis, a convincing argument must be made in a multi-stakeholder dialogue to garner support for the required sustainability transformation.

Once an enabling environment is established, the implementation of the sustainable transformation can begin. Depending on agreed synergies and approaches, material, energy, and information flows between the affected stakeholders should be established (e.g. waste streams from one industry used as a resource in another one; dismantling or repair information flows). It is crucial that the implementation delivers on the values the stakeholders expect. Otherwise, adjustments or even a new round of multi-stakeholder dialogues are required (see Fig. 9.1).

Similarly, internal and interface stakeholders must be considered. This includes not only employees but also different departments and functions within the organization. Wherever possible, this internal stakeholder engagement and alignment should have taken place first (see also Chapters 3.9 and 9). Interface stakeholders act both internally and externally to the organization (e.g. board, Communications, External Affairs).

An example of sustainable stakeholder relationships enabling cross-sector transformations and creating positive, self-accelerating dynamics towards sustainability are electric vehicles. Once the number of electric vehicles passes a certain threshold and becomes the dominant form of transport (positive tipping point), this reduces the costs of battery technology due to economy of scales. Lower-cost batteries in turn provide essential storage capacities to reinforce the positive tipping point to renewable power, which can trigger another tipping point by producing green ammonia for fertilizers, shipping, and so on [4]. In addition, bi-directional charging of electric vehicles can stabilize electricity grids by charging at times of high electricity supply and returning electricity into the grid at times of high demand.

9.1 Sustainable Employee Relations

Considering sustainability in relationships with employees increases job satisfaction, productivity, loyalty, and identification with the employing organization. It also improves workforce retention as well as employer attraction on the increasingly competitive job market as well as in external sustainability ratings [5]. Ideally, sustainability should play a central role in the organization's culture.

Sustainable management of employer-employee relationships is part of the organization's sustainability management. It acts as evidence for the necessity of individual departments of an organization taking ownership of their contributions to sustainability. This is opposed to a heavy reliance on one central sustainability team or officer (see Sect. 3.9). The main reason for this requirement is that sustainability must be incorporated into Human Resources (HR) processes to be effective. Sustainable management of employee relationships must extend beyond relations with employees/unions although building these relations is an important first step. Unions need to be consulted when integrating sustainability in an organization. In fact, forward thinking unions should be inherent allies in making organizations more sustainable.

Sustainable Human Resource management must

(1) Provide HR solutions that support sustainable changes.
(2) Manage HR sustainably. This includes "bringing humanity back into HR Management" [6].

Sustainable HR management can ensure [7]:

- Good treatment of employees to avoid staff turnover, dissatisfaction, stress at work, and loss of morale.
- Sustainable supply and attraction of future employees through training, talent management, and a good work-life balance.
- Engagement of employees in sustainability in terms of changes to their own behavior and voluntary involvement in activities that benefit the environment and society. This also includes encouraging ethical behavior that ensures the organization can implement sustainability practices.

Sustainable HR provides solutions for transformations and carries out its tasks sustainably, ensuring employee loyalty and engagement as well as employer attractiveness.

9.1.1 Support for Sustainable Work Force

An element of sustainable management of human capital is maintaining and improving human capital. Thus, sustainable HR management addresses a wide range of issues like diversity, equity & inclusion (DEI), renumeration & social security, health & safety, stress management & work-life balance, work flexibility (e.g. % of part time, working from home, working in shifts, etc.), recruitment & retention, employee satisfaction (e.g. based on rating in employee or pulse surveys), employee training & development, employee suggestion schemes, etc.

To hold the organization and its HR team accountable, it is important to measure each of these issues and introduce SMART (Stretch, Measurable, Aligned, Realistic, and Time-bound) objectives for each employee. This is required to support transformation. The measurability allows for an assessment of whether an objective was achieved within the time constraints set. Aligning individual targets with the overall sustainability strategy makes the objective relevant. Objectives need to remain realistic to encourage teams to continue to work on them rather than giving up prematurely. Sustainable SMART HR-related objectives could include: same shares of minorities in management and workforce, 10% reductions in occupational accidents year-on-year, 3% of the workforce being mental health first aiders by the end of the year, closing gender pay gaps, improving employee satisfaction in surveys by 10%, increasing employee training hours by a specific percentage, etc. Personal objectives aimed at improving the sustainability of the organization need to be part of SMART objectives of all employees and managers and should align with the organization's and/or department's sustainability goals. These objectives can be formulated by HR and disseminated to all departments, be part of normal performance reviews, and can be linked to bonus and other renumeration programs.

The HR-related objectives drive programs such as:

- Diversity, Equality, and Inclusion (DEI) programs. These include mandatory training, targeted development reviews, career counseling, and talent management etc. They often emphasize gender balances by promoting women to leadership positions, closing gender pay gaps, and offering sensitivity training to combat hidden biases against underrepresented or minority groups within the organization. Inclusion programs for people with special needs, different religions, ethnicities, and sexual orientations exist as well. They support these specific employee groups. It is crucial that DEI programs are an integral part of the organization's culture.
- Concern for employee well-being especially at the psychological level through the provision of counseling services, hotlines for personal advice, and employee assistance program services by external consultants. Classical corporate health services and occupational safety controls should also be implemented. Health and safety programs became crucial during the COVID Pandemic with the workplace offering vaccinations and personal protection equipment (PPE). Programs against workplace bullying and harassment include awareness training, well communicated and managed grievance mechanisms, support for victims, and the penalization of violators [7].
- Flexible work arrangements allow a work-life balance and enable working from distant places, reducing commuting stresses and lost time. Training for management regarding leading remote teams inclusively in such a way that maintains and furthers the team spirit as well as psychological support for remote employees are common HR programs. Such practices do not only support the existing work force but also attract new and external talent. Especially younger office talent is not looking for 9-to-5-office jobs but highly flexible arrangements that blend in with their lives and potentially do not even require them to leave their homes/countries.
- Transformation programs support re- and up-skilling to improve employees' skill sets in terms of decarbonization, digitalization etc. Additionally, sensitive and respectful outplacing programs exist for those employees without longer-term perspectives. These include the organization of internal and external job fairs, paying for interview training or consultancy for occupational re-orientation, etc.
- Employee engagement programs that encourage and support employee engagement with sustainability (see below).

> Sustainability must be included in personal SMART objectives and performance reviews. These can be accompanied by programs for DEI, transformation, and work flexibility.

HR support for sustainable workforces also includes classical tools that ensure succession plans and talent development as well as attract external talents through specific programs and packages.

9.1.2 Employee Engagement for Sustainability

Traditionally, organizations have seen social responsibility as an important building block in their business. Many rely on engagement from their employees to achieve social responsibility goals. As HR teams automatically focus on individuals, HR can play an instrumental role in engaging employees in sustainability. This involves a shift of HR activities from previous Corporate Social Responsibility (CSR) approaches to a focus on the people pillar (CSR looking at profit, people, planet).

Many organizations (large, small, and medium) feel the need to assist their employees in facing their social and environmental problems and encourage them to get involved in volunteer work. This work is mainly conducted outside the company for the benefit of society but can also be done within the company for the benefit of co-workers. Corporate volunteering is a traditional tool in which companies allow employees to use paid work time for community work. The Covid pandemic experience showing the importance of social work beyond the own career supported the understanding of the volunteering value [8].

Employers can provide incentives to employees to improve their personal sustainability footprint by, for example:

- Offering flexible work from home arrangements (reduces commuting emissions) and offering an incentive to switch to renewable electricity providers to make the use of electricity at home cleaner.
- Offering company bikes with preferential leasing rates and tax conditions.
- Offering public transport tickets.
- Offering paid hours/days for social/community/environmental work. This is a win–win approach because it improves employee morale and identification of employees with their own organization and the external community values this.
- Introducing awards, employee suggestion, and incentive schemes for sustainability.
- Introducing competitive sustainability apps in which employees can track their carbon footprint and compete individually and/or as teams for the lowest footprint.
- Establishing employee support groups that collaborate to improve sustainability at the work place, in the community, and at home.
- Offering training and guidance for sustainability from internal and external sustainability experts.

> Organizations support employee engagement in sustainability by offering commuting options, allowing the use of paid working time as community work or introducing gamification apps.

Employee engagement is also key to enabling transformational changes of an organization towards sustainability. It is paramount that the entire organization, including middle-management and employees, fully understands the need for change. Middle-management is often too focused on short-term objectives and may foment resistance to changes that provide long-term benefits unless their objectives are fully aligned with the transformation. Employee reactions to any change usually follow specific patterns [9]:

- Move—aggressively and energetically implementing the change. This is helpful as long as employees remain targeted in their approach and have the required information and support.
- Minimize—supporting the transformation but trying to reduce the required effort as much as possible. This may lead to an incomplete or failed transformation.
- Wait and see—this is a typical reaction of longer-serving employees who have already experienced many proposed changes in their organization that ultimately failed.
- Resist—this can be based on various strategies like raising questions/doubts that undermine team morale, confusing by investigating alternatives, etc.
- Quit—open, explicit opposition and refusal to participate in the change. This may be less harmful than the resist approach.

(Perceived) sudden, unprepared transformations especially trigger fight-or-flight responses. Therefore, sustainable employee relationships must be based on transparency and openness along the whole process, long before a change happens. Organization's leaders must regularly discuss trends and potential disruptions. Disruptions should be embraced, inviting ideas for innovation and turning disruptions into new opportunities. When a disruption occurs, the organization must provide clarity on the details of and reasons for the transformation as well as the impacts of the disruption on the individual employee.

> Sustainable transformations require engagement of employees to ensure the whole organization is on board. Early, open and clear information and dialogue supports this goal.

9.2 Sustainable Finance

The financial sector plays a crucial role in ensuring global sustainable development. Investors are key in directing organizations in the right directions by financing the transformation. Without investments, transformation will not be possible. Similarly, continued investments in unsustainable businesses (like the fossil fuel industry) delay the required transformation and make investors personally responsible for further deteriorations of sustainability in terms of further exceedance of planetary boundaries or continued human rights violations.

9.2.1 Examples for Regulatory Requirements

Several regulations try to steer investments towards more sustainable businesses and make sure that investors have sufficient information to differentiate between sustainable and unsustainable investments (Table 9.1). For the European Union, three pillars form the sustainable finance framework [10]:

- At the center, EU Taxonomy determines the alignment of investment options to what the EU considers sustainable.
- Disclosure requirements provide transparency for market actors allowing them to make sustainable investment decisions. These include Non-Financial Reporting

Table 9.1 Sustainable finance regulations (examples)

	Regulation	Main requirements
US	SEC Climate Guidance	Companies have to make greenhouse gas accounting transparent, provide risk analyses and information on mitigating activities
CA	California Senate Bill 261	Companies in excess of $500 million that do business in California have to prepare biennial climate-related financial risk reports, with the first report due by January 1st, 2026
EU	Taxonomy regulation	Yearly report on taxonomy alignment i.e. assessment whether the manufacturing of a product meets specific criteria, e.g. regarding climate change, pollution prevention, biodiversity, circular economy, and minimum safeguards
EU	Carbon Border Adjustment Mechanism (CBAM)	Import of goods like steel coils, aluminum, etc. will be charged based on their carbon footprint (default values depending on country of production) when being imported from countries without carbon pricing
EU/ UK	ESG Rating Regulation (draft)	Requires ESG rating providers to enhance the integrity, transparency, responsibility, good governance, and independence of ESG rating activities
EU	Green Bond regulation	Establishes requirements for issuers, a registration system, and supervisory framework for external reviewers of European green bonds. Open to all EU and non-EU issuers. Funds raised by the bond must be allocated fully to projects aligned with EU Taxonomy
UK	Climate-related Financial Disclosure	Yearly UK specific reporting based on the Taskforce on Climate-related Financial Disclosures (TCFD) recommendations
UK	Planned UK Green Taxonomy	Expected to build on EU Taxonomy requirements with some UK specific revisions, report expected to be voluntary initially

@ Wulf-Peter Schmidt 2024. All Rights Reserved

Directive (NFRD), Corporate Sustainability Reporting Directive (CSRD) and Sustainable Finance Disclosure Regulation (SFDR).
- Financial Tools—like the EU Green Bond Standard (EU GBS) and EU Climate Benchmarks Regulation—offer sustainable solutions for financial market actors and limit greenwashing in the financial sector.

The California Senate Bill 261 broadly defines "climate-related financial risk" as "material risk of harm to immediate and long-term financial outcomes due to physical and transition risks, including, but not limited to, risks to corporate operations, provision of goods and services, supply chains, employee health and safety, capital and financial investments, institutional investments, financial standing of loan recipients and borrowers, shareholder value, consumer demand, and financial markets and economic health". The reporting requirement is in line with the disclosure framework recommended by the Task Force on Climate-related Financial Disclosures in 2017.

EU Taxonomy is a regulatory framework that determines whether an economic activity contributes positively to the six environmental objectives of the regulation:

- Climate change mitigation (Article 10)—reducing greenhouse gasses in line with climate goals.
- Climate change adaptation (Article 11)—reducing the adverse impact of climate change.
- Water/marine protection (Article 12)—achieving a status of good quality in bodies of water.
- Circular economy (Article 13)—using natural resources in production more efficiently.
- Pollution prevention/control (Article 14)—reducing air, water, land pollution.
- Biodiversity (Article 15)—protecting, conserving, or restoring biodiversity and ecosystems.

In addition, there are four conditions that must be met for an economic activity to be considered environmentally sustainable:

- Substantial contribution to at least one of the six environmental objectives above (Article 9).
- Do no significant harm to the other five environmental objectives (Article 17).
- Compliance with minimum safeguards (Article 18).
- Compliance with technical screening criteria (Article 19 plus delegated acts).

The stringency of these criteria is softened by allowing two other forms of activities to be considered as aligned to the taxonomy. Firstly, activities enabling another economic activity to contribute to one of the six objectives (Article 18). These align with the taxonomy if the "enabling activity" does not lead to a lock-in of assets that undermines long-term environmental goals and provided the enabled activity has a substantial positive environmental impact. Secondly, so-called "transitional activities" (Article 10 (2)) are aligned to the taxonomy. In areas where no alternative, low-carbon solution exists for an economic activity, transitional activities align with the taxonomy under the conditions that they lower greenhouse gas emission levels significantly below the average of the sector or industry, do not prevent the development or deployment of low-carbon alternatives and do not lead to a lock-in of carbon-intensive assets.

EU Taxonomy classifies activities as sustainable that contribute, enable, or are transitionally helpful for at least one of six environmental areas and fulfill further conditions.

The EU Taxonomy regulation includes disclosure requirements for financial market participants as well as large public interest entities. Financial market participants must disclose details of the environmental objective(s) the investment contributes to. This includes disclosing the percentage of investments originating from them. Additionally, transparency regarding non-sustainable company assets and activities is required. Large public interest entities must, for example, disclose the proportion of company turnover, capital expenditure, and operating expenditure associated with sustainable economic activities.

9.2.2 Sustainability Pricing—Example of Carbon Pricing

Pricing carbon or other sustainability items is an important method to financially disincentivize unsustainable practices that may otherwise be cheaper for the actor/polluter in the short term. Introducing carbon pricing into business equations has the potential to bring an end to polluting activities without renumeration (tackling the free-rider issue of polluting for free). Globally, the World Bank identified 75 carbon pricing initiatives in the year 2024 that covered 12.8 Gt CO_2 equivalent which amounted to around 24% of global Greenhouse Gas Emissions [11]. The World Bank defines carbon pricing as "an instrument that captures the external costs of greenhouse gas (GHG) emissions—i.e. the costs of emissions that the public pays for, such as damage to crops, health care costs from heat waves and droughts, and loss of property from flooding and sea level rise—and ties them to their polluter sources through a price, usually in the form of a price on the carbon dioxide (CO_2) emitted. A price on carbon helps shift the burden for the damage from GHG emissions back to those who are responsible for it and who can avoid it. Instead of dictating who should reduce emissions where and how, a carbon price provides an economic signal to emitters, and allows them to decide to either transform their activities and lower their emissions, or continue emitting and paying for their emissions. In this way, the overall environmental goal is supported in the most flexible and least-cost way to society. Placing an adequate price on GHG emissions is of fundamental importance to internalize the external cost of climate change in the broadest possible range of economic decision-making and in setting economic incentives for clean development". An adequate price is best established if there is a functioning, free market.

The biggest market for carbon pricing is the EU Emission Trading Schemes (ETS). It is the commonly named example for carbon pricing and follows the polluter-pays principle. Individual member states like The Netherlands, Sweden, Spain, Romania, or Germany and non-member states like the UK have national ETS systems. Even on local levels, ETS exist (e.g. the Beijing pilot ETS trial). For each ton of CO_2 emitted by industries, energy providers, aviation, etc. within Europe, polluters must purchase a certificate on the market. The CO_2 price was very cheap initially due to the overallocation of free certificates but is rising in the current phase (2021–2030) to a range of between 60 and 100€/ton CO_2. It is important to ensure that there are not more CO_2 certificates on the market than are allowed for a 1.5 °C pathway in support of the Paris Climate Accord and that there is stability in the yearly reductions of available ETS certificates (in EU 2.2% per year which is probably insufficient, 5% in the UK). If this limit is enacted, ETS will achieve the objective of combatting GHG emission at the lowest possible cost to society.

A functioning market with a limited number of emission allowances in line with the 1.5 °C pathway leads to the lowest possible cost in achieving the target.

Similar systems exist for other Greenhouse gasses. The EU Fluorinated Gases Directive requires certain legal entities to buy "quotas" (certificates) reflecting the Global Warming Potential of these F-Gases that are, for instance, used in Air Conditioning Systems.

National systems have their own scope and pricing. For example, the German scheme considers heating oils, natural gas, petrol, and diesel that are not covered by the EU ETS system initially. The pricing is fixed by regulation and will increase from 25€ in 2021 to 55€ per ton CO_2 by 2025 unless the regulator intervenes. This fixed pricing system is less effective because it is an arbitrary pricing level that reflects political will and therefore is not the most efficient marginal abatement cost.

Another carbon pricing approach is a carbon tax. The disadvantage is that the price is not regulated by demand and supply (like for an ideal ETS) but by a regulator that assumes a certain price is sufficiently disincentivizing greenhouse gas emissions. Such carbon taxes exist, for example, in California, several European countries like Sweden, The Netherlands, France, etc., Japan, Korea, and many other countries. The pricing varies but a good example is The Netherlands. They provide planning certainty by increasing the price from 30€ (2021) to €125 per ton CO_2 (2030).

Some companies also use internal carbon prices to check internal business cases and charge an internal levy on their own carbon emissions (e.g. 100$/ton [12]).

9.2.3 LCA-Based Taxation, Import Duties, and Incentives

Starting in 2024, a simplified cradle-to-gate LCA performance assessment is an eligibility criterion for purchase incentives for electric vehicles (EVs) in France [13]. Carbon footprint values for the production of steel, aluminum, other materials, the battery, assembly as well as vehicle transportation (to France) based on default data from the region or country of origin are added up (suggesting systematically higher carbon footprints for China than for Europe or US). If manufacturers have primary data from their specific supplier/production they need to validate this data.

The calculation of the Ecobonus is a simplified cradle-to-gate carbon footprint approach based on the following equation

$$EC_{Version} = EC_{ferrous} + EC_{aluminium} + EC_{AM} + EC_{battery} + EC_{ATI} + EC_{transport} \tag{9.1}$$

where:

$EC_{ferrous}$ is the carbon footprint of the production of the ferrous metals consumed for the manufacturing of the vehicle type, excluding batteries, expressed in kilograms of CO_2 equivalent (kg-eq CO_2);
$EC_{aluminum}$ is the carbon footprint of the aluminum (pure and alloyed) consumed for the manufacture of the vehicle version, excluding batteries, expressed in kg-eq CO_2;
EC_{AM} is the carbon footprint of the production of materials, other than ferrous metals, and aluminum (pure and alloyed), used for the manufacturing of the vehicle, excluding batteries, expressed in kg-eq CO_2;
$EC_{battery}$ is the carbon footprint related to the production of the battery, expressed in kg-eq CO_2;
EC_{ATI} is the carbon footprint related to the energy required for intermediate transformations and assembly of the vehicle, excluding batteries, expressed in kg-eq CO_2;
$EC_{transport}$ is the carbon footprint related to the transport of the vehicle in question from its assembly site to its distribution site in France, expressed in kg-eq CO_2.

The intention of this approach is to focus the EV purchase incentive on EU produced smaller EVs. However, this approach is an example for a protectionist (mis)use of the scientific life cycle thinking. It uses untransparent default data, arbitrary factors, and thresholds as well as questionable system boundaries that for example do not support the introduction of technologies or materials that are linked to high-carbon emissions in production but significantly reduce carbon emissions during the use and/or recycling phase.

Similarly, Regulation (EU) 2023/956 establishes a carbon border adjustment mechanism (CBAM) to address the carbon footprint (cradle-to-gate) of certain goods (iron and steel, cement, aluminum, fertilizers, hydrogen, and electricity) if they are imported into the customs union. It addresses the import of goods from countries that do not price carbon. The intention is to establish a fair level playing field and avoid unfair competition between imports that have been produced without being charged for the CO_2 emissions and those produced domestically and covered by ETS. CBAM requires certified, specific primary data as well as World Trade Organization (WTO) acceptance.

LCA-based taxation, import duties, and incentives are currently only based on cradle-to-gate considerations. Issues include data availability, Life Cycle and political trade-offs.

9.2.4 Insurance and Investor Approaches for Sustainability

Insurers are already heavily affected by high costs linked to the exceedance of planetary boundaries. Only considering weather events in 2022 that were intensified by climate change, direct economic cost of $360 billion arose of which around 40% was covered by insurance providers [14]. Thus, the CEO of an insurer stated "A + 4 °C world is not insurable. Unsustainable business will become uninvestable and uninsurable" [15].

Insurance companies can play a crucial role in driving sustainability. If insurance premiums are high enough for unsustainable production, companies move towards sustainability because premiums undermine even short-term business cases for unsustainable production methods [16]. Ideally, unsustainable practices should not be insured at all to further increase the pressure on companies. Usually, companies discontinue their activities if there is no insurance. The loss of business for insurers is mitigated by an increase in the credibility of insurer's ESG claims and image and the reduced business risks for insurers.

Similarly, this encourages investors to steer their investments away from unsustainable to more sustainable investments. Again, companies not raising required investments due to unsustainable business models will have to change their business models quickly. More sustainable businesses received investments totaling to an estimated $30.3 trillions in 2022 [17]. This is mutually beneficial to both, companies and investors, as "research has discovered that environmentally responsible practices not only enhance shareholder value but also the value accrued to nonfinancial stakeholders" [18]. The same applies for ESG responsible practices [19]. There was a short reversal of this situation due to the soaring energy prices after the start of the Russian war against Ukraine that created additional profits for some fossil fuel industries. However, this was only a temporary exemption from the rule.

The finance sector, insurers & investors determine whether unsustainable businesses continue with their detrimental activities or whether more sustainable transformations are possible.

EU Taxonomy regulation is meant to ensure that investors are supplied with standardized information and are not misled in terms of the alignment of the investment with what is seen as sustainable under EU rules (Table 9.1). Similarly, individual guidance on sustainability is provided by various rating agencies [20–22].

9.3 Sustainable Policy Making, Judiciary and Advocacy

Sustainable policy making and advocacy should be rooted in scientific evidence, in line with human rights, support the UN sustainable development goals and prioritize the most pressing sustainability issues. These issues include climate change and biodiversity in particular. The challenge is to act decisively and with urgency while "leaving no one behind". The EU Green deal promises this [23]. Given the urgency for action, it is important to ensure that all life cycle stakeholders act. Policy and legislation are key in pushing all stakeholders in the right direction. It is insufficient to focus only on technology or behavioral changes of consumers and citizens. It is not an "either or" but an "all hands-on deck" situation. All stakeholders have to transform. Waiting for the perfect solution is insufficient as this solution will arrive too late. The perfect is the enemy of the good.

A particular responsibility in supporting the sustainable transformation falls to life cycle stakeholder advocacy representatives (lobbyists). They must have an embracing, constructive "yes and" attitude: "yes" we support the required transformation "and" we think the following enabling policy will accelerate this transformation. This demonstrates an engaged, "can-do" attitude and focusses on making the inevitable change happen, rather than focusing resources on finding reasons to not change. This differs from the tactical "yes but" policy that implies a theoretical support for sustainability "but" only under certain conditions (subsidies, later timing, softening of targets). Such an attitude shows that lobbyists still have not fully understood or accepted the need to transform if they wish to remain in business. It is also opposed to the negative "no" attitude that represents denial.

These three advocacy options can be positioned on the change curve that was originally developed in the 1960s by Elisabeth Kubler-Ross. It is widely used as a method to help understand the reactions of people to significant changes and upheavals (Fig. 9.2). The first reaction to a significant change or required disruption/transformation is usually shock. This initial shock, while frequently short-lived, can result in a temporary loss of productivity and performance as well as an increased

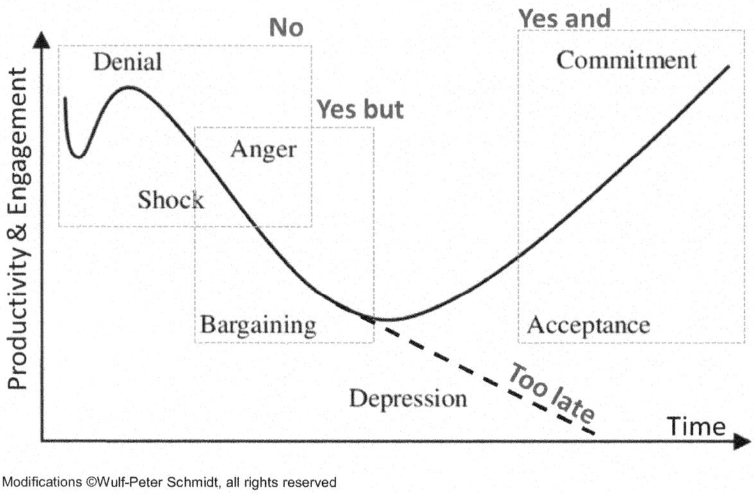

Fig. 9.2 Advocacy options in the Kübler-Ross Change Curve (modified [24])

desire for guidance and reassurance (e.g. citizens asking for strong leaders). This shock is often a result of:

- A lack of information.
- Fear of the unknown.
- Fear of looking stupid or doing something wrong.

Denial by individuals and organizations often follows this initial shock. They try to stay in the past by focusing on the past instead of future demands. The past is glorified to deny any need for change and any signs that the change is happening (e.g. climate change) are typically ignored/denied/rationalized away. Performance often returns to pre-shock levels, business as usual is prioritized and thus parties see no need for any forward planning for the potential change. If there is no clear and constant communication and explanation for the required change, organizations and individuals continue as to say "no" to the transformation as long as possible.

After shock and denial, anger is often the next stage. Anger is commonly focused on those that continue to communicate, advocate, or actually implement the change. If these communicators/advocators are strong, organizations and individuals try to negotiate from the position that change is not needed.

This often coincides with a tactical modification to the "no to any change" position. Now, organizations or individuals suggest they could consider the change if there are strong incentives (subsidies, softening of change) or even more insidiously if technology neutrality that allows for compromises in performance and loop holes in regulation (e.g. eFuels instead of Zero Tailpipe Emission Vehicles). This is termed the "yes but" position. This stage reveals feelings like suspicion, skepticism, and frustration. The risk is that policy makers give in, leading to a partial transformation that is insufficient in addressing the issues. In turn, this means another round of changes is required—causing even more frustrations for all life cycle stakeholders.

This cycle continues until the lowest point of the curve is reached. Here, the anger and illusion of bargaining begins to wear off and organizations and individuals realize that the full change is inevitable. Anxiety levels peak, morale is low and the decisive question is whether constructive self-doubt leads to positive changes fast enough or whether negativity and destructive behaviors continue until the organizations and individuals have crossed the point of no return and hit a metaphorical wall (bankruptcy, extinction, etc.)—see the dotted line in Fig. 9.1. In this stage, there is often a tendency to focus on day-to-day tasks or other crises to distract from the wider picture.

If this lowest point is passed positively and constructively, acceptance that change is inevitable is reached, leading to more optimistic and enthusiastic moods. Organizations and individuals begin to work with the changes rather than against them. The focus is firmly on the future and making the change happen ("yes and"). The final step creates a sense that real progress can be and is now being made. Primary feelings now include acceptance, hope, and trust. In the beginning, productivity remains low until the momentum of change convinces everybody. Regular communication and progress reports remain key.

> Advocacy focused on making a success out of the required change to more sustainable business models helps the own organization the most.

An example of a "yes and" advocacy is the EU regulation for 100% CO_2 reduction for new registered passenger cars and light commercial vehicles. Some automotive manufacturers (Ford, Volvo) issued an open call to EU policy makers in favor of the effectively proposed ban of new petrol and diesel vehicles by 2035 [25]. This is especially remarkable because these automotive manufacturers do not exclusively produce EVs which would make such a position easy to take. As a matter of fact, these manufacturers predominantly produced petrol and diesel vehicles when they started to advocate for a ban of these vehicles. This is an example of companies understanding the need for transformation, following their "north star" (Sect. 3.7), doing the right thing, and disrupting themselves. They committed externally to a change, wanting to ensure that regulators did not lag behind industrial realities but provide the right regulatory framework. This helped them gain advocacy credibility externally and ensured alignment with sustainability internally.

Judicial systems are similarly important in driving (or preventing) sustainability. The State of California sued several oil companies—Exxon Mobil, Shell, BP, ConocoPhillips, and Chevron—and the industry trade group American Petroleum Institute (API) in September 2023, claiming that their actions have caused tens of billions of dollars in damage and that they deceived the public by downplaying the risks posed by fossil fuels. The State of California sought the creation of an abatement fund to pay for the future damages caused by climate-related disasters in the state. Similarly, in 2021, a Dutch court ordered the Royal Dutch Shell (RDS) to reduce the net worldwide CO_2 emissions of the Shell group compared to 2019 by 45% by 2030 to support the 1.5 °C target. This ruling followed a claim initiated by seven foundations and associations as well as over 17,000 individual claimants [26] and has been appealed by Shell.

References

1. R. Tapaninaho, A. Heikkinen, Value creation in circular economy business for sustainability: a stakeholder relationship perspective. Bus. Strat. Environ. **31**, 2728–2740 (2022)
2. J. Mayers, *Stakeholder power analysis* (International Institute for Environment and Development, 2005).
3. T. M. Lenton, L. Laybourn, D. I. Armstrong McKay, S. Loriani, J. F. Abrams, S. J. Lade, J. F. Donges, M. Milkoreit, S. R. Smith, E. Bailey, T. Powell, C. Z. C. B. J. B. J. D. Fesenfeld, *Global Tipping Points Report 2023* (University of Exeter, Exeter, 2023).
4. C. Lyona, D. Cordell, B. Jacobs, J. Martin-Ortega, R. Marshall, M.A. Camargo-Valero, E. Sherry, Five pillars for stakeholder analyses in sustainability transformations: The global case of phosphorus. Environ. Sci. Policy **107**, 80–89 (2020)
5. R. Sitko, *Sustainable Human Resource Management—Using HRM to achieve long-term social, environmental and business goals* (2023).
6. F. J. Gracia, Special Issue "Sustainability and Human Resources Management, *Sustainability* (2023).
7. N. Papalexandris, Sustainable development and the critical role of HRM. Stud. Univ. Babes-Bolyai Oecon. **67**(3), 27–36 (2022)
8. K. C. Sahoo, M. R. Sahay, S. Dubey, S. Nayak, S. Negi, P. Mahapatra, D. Bhattacharya, M. O. Del Barrio, S. Pati, Community engagement and involvement in managing the COVID-19 pandemic among urban poor in low-and middle-income countries: a systematic scoping review and stakeholders mapping. Glob. Health Action **16**(1) (2023)
9. C. Bateman, M. Pleshette, A. Cindrich, C. Phillips, *Change—how to turn uncertainty into opportunity* (Mango Publishing, 2023).
10. M. Littlehale, EU sustainable finance: a growing taxonomy; Politicopro Analysis, 31st of July 2023.
11. World Bank, *Carbon pricing dashboard* 2024, https://carbonpricingdashboard.worldbank.org/
12. J. Guo, D. Kubli, P. Saner, The economics of climate change: no action not an option (Swiss Re Institute, Zurich, 2021).
13. Ministère de la Transition énergétique République Francaise, Arrêté du 19 septembre 2023 relatif à la méthodologie de calcul du score environnemental et à la valeur de score minimale à atteindre pour l'éligibilité au bonus écologique pour les voitures particulières neuves électriques. J. Off. République Française (2023, 20 Sep.).
14. Reuters, *Less than half of global cost of climate disasters insured, broker Gallagher Re says* (2023).
15. T. Buberl, "ESG: A growing sense of urgency," Quote of CEO of AXA from UN Environment Program (UNEP) One Planet Summit, 2017.
16. I. Colivicchi, G. Iannucci, The environmental responsibility of firms and insurance coverage in an evolutionary game. Dyn. Games Appl. **3**, 801–818 (2023)
17. Global Sustainable Investment Alliance, *Global sustainable investment review 2022* (2023).
18. D. Gilchrist, J. Yu, R. Zhong, The limits of green finance: a survey of literature in the context of green bonds and green loans. Sustainability **13**(2), 478 (2021)

19. U. Saeed Bhutta, A. Tariq, M. Farrukh, A. Raza, M. K. Iqbal, Green bonds for sustainable development: review of literature on development and impact of green bonds. Technol. Forecast. Soc. Chang. **175** (2022).
20. S&P Dow Jones Indices, *Dow Jones Sustainability World Index* [Online]. https://www.spglobal.com/spdji/en/indices/esg/dow-jones-sustainability-world-index/#overview
21. MSCI, MSCI [Online]. https://www.msci.com/
22. Sustainalytics, Morningstar sustainability rating for funds [Online]. https://www.sustainalytics.com/investor-solutions/analytic-reporting-solutions/morningstar-sustainability-rating-for-funds
23. EU Commission, *The European green deal—communication* (Brussels, 2019).
24. E. Kübler-Ross, *On death and dying* (Macmillan, New York, 1969)
25. A call to the European Parliament and EU Governments to support the European Commision's proposed phase-out of new diesel and petrol cars by 2035, 17 May 2022. [Online]. https://www.theclimategroup.org/call-european-parliament-and-eu-governments-support-european-commissions-proposed-phase-out-new.
26. Royal Dutch Shell must reduce CO_2 emissions (Den Haag, 2023).

Chapter 10
Technical Sustainability Management: Challenges and Solutions

Abstract The book details different aspects of Technical Sustainability Management that combines management, good governance, and engineering based on life cycle thinking to improve the relevant social, environmental, and economic aspects holistically. This chapter analyzes the most important challenges to sustainability progress and Technical Sustainability Management. In particular, it considers drawbacks in the areas of resources, technology, human aspects, information, legacy, and crises. Based on this, potential solutions for these challenges are provided, mainly through the lens of the Technical Sustainability Management toolbox.

Keywords Sustainability · Drawbacks · Challenges · Solutions · Change management · Crises · Human behavior

This book started by explaining the sustainable development goals and the challenges humanity faced within its attempt to achieve these goals. These challenges consist of a multitude of parallel crises such as

- Climate change, species mass extinction, endangered biodiversity, and many other issues leading to the exceedance of multiple planetary boundaries that must not be crossed if a "safe operating space for humanity" is to be maintained.
- Social intragenerational injustice and inequalities leading to poverty, hunger, social unrest, migration, etc.
- Economic intergenerational injustice and monopolies leading to reduced opportunities and negative impacts on other sustainability dimensions.
- Corruption that weakens democracies and exacerbates inequalities further.

Against this rather depressing backdrop, a promising set of solutions is available. Key to tackling these challenges is a holistic perspective that ensures unintended side-effects and trade-offs of solutions are identified from the beginning to allow for informed decision as well as mitigating actions. Hence, environmental, social, and economic life cycle tools such as Life Cycle Assessment (LCA), Social LCA, and Life Cycle Costing form the fundamental basis for all solutions and approaches considered (Chapter 2).

© The Author(s), under exclusive license to Springer Nature Switzerland AG 2024

W. Schmidt, *Solutions For Sustainability Challenges*,
https://doi.org/10.1007/978-3-031-63624-0_10

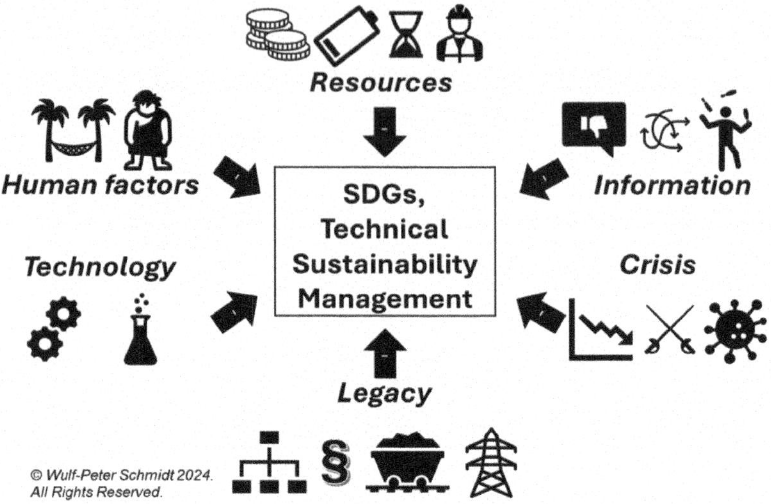

Fig. 10.1 Drawbacks and challenges for SDGs and technical sustainability management

Profound and quick transformational changes are crucial, given the multitude of sustainability issues and the urgency of these challenges. Holistically redesigning business strategies and models as well as organizational policies ensures such changes occur. Sustainability strategies need to be an integral part of the organization's overall strategy and should not be a stand-alone approach. Integrating sustainability aspects in classical and modern strategy development and (change) management tools is crucial (Chapter 3). The same applies to the integration of sustainability in product development because this step determines the potential product impacts along the entire life cycle. Through this, the book conveys opportunities for all strategists and product designers to support the global effort towards more sustainable development. The following chapters offer various suggestions to anyone involved in purchasing, transportation, manufacturing, services, the circular economy, HR, IT, communications, finance, policy or lobbying. The decisive impact of consumers is also outlined, offering opportunities for each consumer to positively contribute to sustainability.

The solutions outlined in this book demonstrate the importance of combining sustainable management and governance practices with sustainable technologies and applying a holistic life cycle approach to improve social, environmental, and economic dimensions—i.e. Technical Sustainability Management. Of course, there are several difficulties and drawbacks in all three aspects (management, technology, life cycle thinking). These difficulties can be categorizes as issues with resources, human aspects, information, technology, legacy, and crises. Each challenge delays progress towards the SDGs and acts as a potential drawback of Technical Sustainability Management (see Fig. 10.1).

Each of these challenges can be addressed by previously mentioned Technical Sustainability Management approaches based on holistic life cycle thinking.

10.1 Resource Challenges and Solutions

Resource challenges include (large) added costs of more sustainable options for the final customer and/or the investing organization. In a global survey of companies, around a fifth of participants named high costs for more sustainable solutions as a major issue [1]. This can be linked to high short-term return-on-invest (RoI) expectations/standards and/or lack of cash or financing options. For example, major companies expect the RoI to be above alternative investment RoI's i.e. above 10–15% (i.e. latest within 10 years) while even the lowest cost investment for new electricity generation—wind power (see Fig. 6.5)—has an RoI between 12 to 20 years [2]. Raw material availability is often an issue in areas where the current mining and material generation capacity is low compared to the projected demand for new, more sustainable technologies—in particular the so-called Critical Raw Materials (defined by the EU Commission) like Lithium and Cobalt that are needed for electrification in almost all sectors [3]. Equally, the availability of low-carbon liquid and gaseous energy carriers like biogas (to substitute natural gas) is limited in several regions (see Sect. 6.6) which restricts the opportunities to decarbonize production in the short term [4].

Similarly, the availability of employees or service providers with the necessary skill set for more sustainable technologies or for establishing the required infrastructure is limited, in particular in regions with rapidly aging work forces (Europe).

Last but not least, time constraints can be a reason not to pursue more sustainable solutions. For example, reusing previous designs is possible in a fraction of the time that it takes to design a completely new product or service that has a better sustainability performance. Also, the time it takes for better solutions to be available/implementable might be too long.

> Resource limitations around finances, time, raw materials, energy, and work forces can be addressed by sustainable financing, design, and other Technical Sustainability Management tools.

While Technical Sustainability Management can be negatively impacted by these resource limitations, it provides potential solution at the same time:

- Financial issues can be addressed by sustainable finance (see Sect. 9.2). This includes financing (only) sustainable investments and adding costs for non-sustainable investments or consumption (e.g. CO_2 taxation or emission trading schemes) to close the cost gap. In addition, organizations must adjust their RoI requirements and extend their understanding of business cases (see Sects. 4.6.5 and 9.2.2). Also, on-costs can often be put into perspective—e.g. by considering the life cycle perspective and/or environmental business cases (Sect. 4.6.5).

- Raw material scarcity can be addressed by design for recycling or low-impact resources (see also Sect. 4.1.1), backstop technologies (e.g. substituting rare earth materials in magnets [5]) and circular economy approaches (see also Chapter 8).
- The lack of skilled work force and service providers is best addressed by education (see Sect. 7.6) including also up-skilling existing work forces from declining industries that need new job opportunities for their employees.
- Time constraints can be mitigated by developing long-term strategies and planning that ensures sufficient lead-times for changes. They may be based on long-term futuring and scenario planning (see Sect. 3.4). Of course, urgent action is required, so establishing plans that have sufficient buffer and lead-time rather than relying on a "hockey-stick" approach in which everything is planned for the latest possible point in time is warranted.

10.2 Human Challenges and Solutions

Several major difficulties in establishing Technical Sustainability Management relate to human nature. Of course, human nature and behavior is unique and regarded as the result of cultural shaping (cultural artifacts, education, socialization etc.), genetic predisposition (pattern recognition based on instincts, needs, drives, etc.), and situational correctives [6].

However, there are legacy values and cultures that impact every human in some way and some of these can directly conflict with solutions for sustainability as well as Technical Sustainability Management. In particular, cultures and lifestyles that value intensive consumption patterns and materialistic wealth as well as short-term and egoistic thinking accelerate the exceedance of planetary boundaries [7].

Similarly, human resistance against (any) change or new approaches is not helpful for the required transformation. This may delay or even prevent progress from being made. Examples include skepticism or even resistance against wind power (while accepting motorways or industrial areas that have larger impacts on the landscape or noise levels) or electric vehicles (questioning environmental credentials despite scientific evidence or criticizing the high prices while accepting them for other vehicles). The typical human response to approaches perceived as sudden, or pushy attempts to introduce changes is sullen resistance (see in Germany political discussions around perceived prohibitions around heating or food).

Another issue with human behavior is its tendency towards irrationality. Current scientific thought differs significantly from historical assumptions about rational human behavior [8]. Another common human behavioral issue relates to short-term thinking. Humans have inherent difficulties with adhering and understanding long-term plans and fall prey to impulsive decision-making that neglects consequences. This might be a result of human brains seeking immediate rewards and avoiding immediate pain [9]. Also, the real business world, often focusses on short-term performance gains while neglecting long-term (business) needs. A recent business consultant study suggested that in a survey of around 200 corporate directors and senior managers around 60% of respondents point to differences in leadership opinions on "how to balance short-term considerations with long-term investments and sustainable growth" needs [10]. While this survey is not necessarily representative on a global scale, it corresponds with the preference of most companies for short-term profits. The same study showed that only 8% of the responding company leaders were completely satisfied with the business rationale of a transformation towards sustainability. This leads to a situation where companies quickly change direction or lose momentum when faced with difficulties—see the revised long-term electrification ambitions of some automotive companies when confronted with a slower than expected uptake of electric vehicles [11]. Short-term thinking comes hand in hand with procrastination, laziness, lack of resilience, unwillingness to take personal responsibility, and a refusal to leave the comfort zone—all of these can undermine the implementation of more sustainable solutions by delaying necessary changes and rejecting anything that looks marginally less comfortable. Short attention spans do not support the necessary understanding of complex and holistic decision criteria of Technical Sustainability Management. Another typical human shortfall is the limited ability to learn from experience and mistakes [9].

Similarly, Technical Sustainability Management can be negatively impacted by these shortfalls of human behaviors. However, it also provides potential solutions:

- Sustainable marketing (see Sect. 7.7) can help direct customer behavior towards more sustainable patterns, establish longer-term relationships, and provide enough information to point customers towards more sustainable options.
- Product development can incorporate insights into human behavior into design solutions. For instance, the more sustainable use could be designed to be the most comfortable and straightforward option to utilize a product (see for examples Sect. 4.4).
- Pricing can incorporate externalities (see for examples in Sect. 9.2.3) and regulations can direct all life cycle stakeholders (e.g. The Netherlands and other EU countries prohibiting new registrations of internal combustion vehicles from 2030), However, regulatory restrictions only stick if they are embedded within cultural shifts which takes time to achieve. A clearly explained north star underpinned by scientific research (Sect. 3.7) supports this process.
- Structured change management that incorporates the perspective of all affected life cycle stakeholders and limits the social impact of the required change/transformation as early as possible can mitigate sullen internal and external resistance (e.g. by letting citizens financially benefit from new wind farms in their neighborhood or making a convincing business case for change to "galvanize people behind the complex enterprise transformation required" [10])—see also Sect. 3.9.

It is important to note that human behavior also has its upsides. Hundreds of experiments in various cultural contextes have shown that short-term selfishness is often overcome by a sense of fairness and reciprocity as well as support for social behaviors [8].

Human limitations in regards to short-term thinking, selfishness, and comfort can be addressed by change management, design, and other Technical Sustainability Management tools.

10.3 Information Challenges and Solutions

(Dis-)information can impact human behavior and undermine Technical Sustainability Management. This can be done on purpose—for example when the fossil fuel industry questions man-made climate change [12] or more recently in media campaigns against electric vehicles, particularly in the US [13] and UK [14] – where very obvious financial, short-term interests motivate disinformation. Less obvious examples are the eight possible greenwashing approaches (see Sect. 7.7.1) of marketing campaigns. Similar approaches outside of traditional marketing like attempts to confuse discussions by bringing up wrong, irrelevant information and questions also exist (e.g. suggesting that a reduced growth rate in EV sales means a decrease in sales which is mathematically incorrect [15]). Ideological debates may fit into this category as well: certain terms like "ESG" are seen as "woke" and are therefore rejected instinctively [16].

On the other hand, a lack of information can prevent companies from obtaining the data required for a thorough analysis that underpins Technical Sustainability Management, in particular its complex and data-intensive life cycle-based tools. Each analyzed process in a life cycle study can require hundreds of data points to be comprehensive. Data for some sustainability issues (like biodiversity, external costs, corruption) are more difficult to gather than others (like energy consumption, direct costs, income level). In particular, small-and-medium organizations or private consumers cannot spend their time looking for all this information.

Data complexity and inconsistencies add to the challenge. Product and sometimes organizational life cycles are spread globally due to global supply chains. Ultimately, almost everything is linked to everything else which can lead to infinitive data needs. Conflicting information and trade-offs can lead to a situation where no decision is taken, effectively delaying progress towards sustainability.

Again, Technical Sustainability Management can be negatively impacted by these information issues but it also provides potential solutions:

- Cut-off criteria and methods that carefully limit the necessary information have been established for all life cycle tools (see for example Sects. 2.2.1.2 and 2.2.2.2.1).
- Education about sustainability can communicate facts clearly and enable current and future employees and decision makers to use the right tools and approaches when assessing the situation and potential options.

- Similarly, sustainable communication can demystify the complexity of sustainability (solutions), correct disinformation and unmask anti-ESG / anti-sustainability campaigns (see for example fact checks of false electric vehicle myths [17]). It is important that this communication is adapted to the perspective of the audience (see Sect. 3.9).
- Sustainability advocacy of leading businesses and companies can help to shift the narrative and establish sustainability as an economic opportunity and necessity rather than a nice-to-have for better times (e.g. well-ESG-rated companies have higher likelihoods of economic success as explained in Sect. 1.6 and transitions of old economies may ensure that these companies still have products that can be sold at all).

Information issues around disinformation, information availability, and data complexity can be addressed by education, streamlining, and other Technical Sustainability Management tools.

10.4 Technology Challenges and Solutions

Major technology challenges in the area of sustainability include availability, maturity, quality, robustness, acceptance, and sustainability trade-offs of technology.

In most areas, technology for sustainability is available in principle (e.g. renewable energy, lower-impact steel, cement, recycling technology—see Chapters 6 and 8). However, maturity of some solutions might be limited if still on a laboratory scale (e.g. electrical recovery of carbon fibers in composites; see Sect. 8.4). The robustness and quality of technology is often linked to the maturity of the technology. Robustness can be an issue if, for example, slight changes of the input material or of process parameters impact output quality. This is often the case with new, advanced technology approaches which limits opportunities to quickly improve these technologies (see for example EV battery production/Sect. 6.3).

The acceptance of technology can be impacted by disinformation (see Sect. 10.3) or actual trade-offs. For example, resource availability or sustainability of the required resources—like Cobalt for batteries—as well as on-costs are often-raised concerns.

Issues regarding technology of course directly impact Technical Sustainability Management, however, there are in-built solutions:

- The maturity and development of new technologies can be supported by sustainable financing. There are several start-up crowdfunding platforms focused on sustainable solutions. Education, research, and innovation to resolve technical limitations are made possible by such finance.

- Dealing with and resolving trade-offs is an integral part of the life cycle decision tools presented in this book—in particular in their interpretation phase (see Sects. 2.2.4, 2.3.4, and 2.4.4) but also directly by comparing savings in one life cycle phase to potentially higher impacts in others (e.g. for electric vehicles showing that the net impact is beneficial, see Fig. 2.2).

Technology issues around availability, maturity, and trade-offs can be addressed by the key Technical Sustainability Management tools of sustainable financing and life cycle thinking.

10.5 Legacy Challenges and Solutions

Infrastructure (like pipelines, buildings), technology (e.g. combustion), and products (like heating systems, vehicles) with long lifetimes typically hold up fast progress because their replacement takes time and their running costs are often low compared to the required investment for replacement. They also might have already been written-off or their existence may reduce the perceived urgency to change. Similarly, legacy investments can force decision makers into old solutions. If a country or company is based on mining, it takes a long time to transform into a sustainable service economy or company.

Legacy organizational structures, processes, and skill sets of employees as well as decision makers that do not incorporate sustainability needs and perspectives slow down the ability of new ideas to be realized as more sustainable, interacting structures.

Legacy regulations that provide (free) rights to pollute, do not protect workers rights or strip down governmental bodies to prevent them from enforcing social, environmental, or fundamental democratic rights are not benefitical to Technical Sustainability Management either.

Nevertheless, Technical Sustainability Management provides potential solutions to mitigate these legacy drawbacks:

- Retrofitting and upfitting might be suitable options to improve the sustainability of existing infrastructure, technologies, and long-living products. For example, existing buildings can be upgraded by improving insulation, adding solar panels etc.
- Change management can help move organizations towards more sustainable business models (see Sect. 3.9) while education can upgrade the skill set of decision makers and employees towards sustainability
- "Yes and" advocacy for more sustainable regulations (see Sect. 9.3) can pave the way for a regulatory environment supportive of sustainable transformations.

Legacy issues around long-living infrastructure or products and regulations can be addressed by upfitting, change management and other Technical Sustainability Management tools.

10.6 Crises—Challenges and Solutions

There is almost always a crisis that overshadows the various sustainability crises and SDG goals for the year 2030. For example, the 2008 global financial crisis terminated a lot of sustainability efforts because companies were struggling to survive financially and focused primarily on cash management, postponing any social or environmental projects. Empirical studies covering 178 countries between 1995 and 2019 [18] have shown that resilience to climate change shocks has increased, however, financial crises lead to short-term deterioration in countries resilience—in particular in developing countries. Economic downturns cause the vulnerability to climate change to increase. Obviously, social sustainability issues worsen in economic downturns. On the other hand, economic downturns tend to decrease economic activity and therefore emissions.

The Covid-pandemic crisis had similar impacts. Any market collapse or economic downturn limits the opportunity of organizations to invest in a more sustainable future.

Actual or even threats of wars distract governments and redirect their spending towards defense rather than sustainable transformations. There also seems to be an overall political crisis leading to an increase in autocratic regimes, globally. While the share of the global population living in liberal democracies has increased from 0% in 1845 to 18.5% in 2022 this is less than the 24.8% that was achieved in 2009 [6].

Media lives from perpetually finding and/or establishing new crises that are immediate and easy to understand rather than constantly reporting on previously explored, complicated, or distant (in space and time) sustainability crises. This distracts the public debate and makes it difficult for society—and therefore politicians—to stay focused on fundamental sustainability items for longer periods of time.

Solutions in this area include:

- Using crises as opportunities for Technical Sustainability Management. For example, the pandemic enabled a cultural shift towards remote working and replaced business trips with virtual or hybrid conferences which reduced the need for air travel. It also increased awareness of supply chain issues, enabling the addition of social aspects to supply chain considerations.
- Sustainable policy making can lock-in demand and emission reductions achieved in a crisis and "ensure that energy pricing, investments, and other policies are

conducive toward innovations that create low-carbon societies (through appropriate investments, either public, private, or mixed through Public Private Partnerships)" [18]. Also, the financial opportunities for countries to continue to provide subsidies for non-sustainable practices are lower and opportunities for taxing non-sustainable products exist.

- All forms of media must constantly communicate sustainability issues, maintaining the sustainability debate in the public's attention.
- Sustainable development is based on partnerships and global cooperation which in itself increases global resilience against crises and in the longer run also reduces the potential for armed conflict as it removes several reasons for military conflicts.
- Most important is the recommendation "never give up"—even if a financial crisis or economic downturn or market collapse stops your sustainability strategy or your more sustainable organization runs out of cash, use this as a learning opportunity for an even better strategy in a more robust organization afterward. There is always an economic up-turn where you can be the first one on stage.

> Global financial, health and conflict crises often end sustainability projects but can be turned into opportunities by locking in reduced demands and emissions or ending non-sustainable subsidies.

10.7 Outlook and Conclusions

While there are lots of reasons why progress in sustainability might not work out, the case has been made that Technical Sustainability Management offers all the solutions required to overcome potential drawbacks (Fig. 10.2).

It is remarkable that all the required technologies and insights necessary to tackle sustainability are available and well-known. Sustainability solutions exist in all sectors, professions, and organizations. Furthermore, fundamental technical solutions that positively impact all sectors such as onshore wind energy are the most economically favorable solutions (Sect. 7.5). In contrast, old technologies that endanger sustainability are becoming increasingly expensive—a trend that is supported by carbon pricing. Similarly, supply chain regulations are increasingly being implemented to tackle human right violations globally. This provides further evidence for the global momentum towards a more sustainable world. In fact, the increasing number of sustainability regulations (see for example Sects. 5.1, 7.7.1, 9.2.1) should not be seen as a burden to be complied with but as an opportunity to establish competitive advantages by exceeding regulations in terms of timing and ambition.

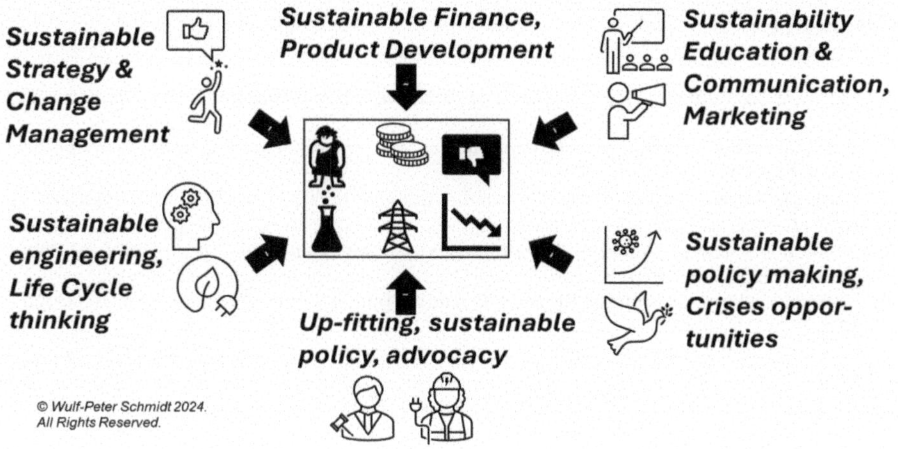

Fig. 10.2 Technical sustainability management solutions for potential drawbacks for sustainability

Additionally, there is statistical evidence that the situation is improving in different areas:

- Many countries have decoupled economic growth from CO_2 emissions i.e. they managed to decrease their total emissions while achieving net growth [19].
- Global growth in clean energy was twice as large as that of fossil fuels from 2019 to 2023 and the outlook is even better [20].
- Global air pollution has decreased so significantly that air pollution-related deaths have decreased by around 45% between 1990 and 2019 [21].
- Vaccines are being developed at a much faster rate than ever, reducing the devastating impact of pandemics.
- Across the world, people live longer—in fact the average life expectancy has more than doubled between 1900 and 2021 from 32 to 71 years [22].
- The share of the global population living in extreme poverty has decreased from more than 75% in 1820 to around 10% in 2018 [23].

There are many more encouraging examples. Each reader is invited and empowered to apply the outlined solutions, opportunities, and tools of Technical Sustainability Management to support and accelerate sustainability transformations. The intention is to democratize sustainability and make sure that sustainability is not just a job for sustainability specialists, the CSO or the CEO. Instead, each position within each organization as well as each life cycle stakeholder is encouraged to actively contribute to the progress towards sustainability.

References

1. McKinsey & Company, *How companies capture the value of sustainability: survey findings* (2021).
2. Whether Guard—Lightning Tech, *Wind turbine cost: the economics of wind energy in 2023* (April 24, 2023).
3. European Comission, *Study on the critical raw materials for the EU 2023—final report*. Brussels.
4. International Energy Agency, *Outlook for biogas and biomethane—prospects for organic growth* (IEA, Paris, 2020)
5. Veekim, Saubere performance für Ihren E-Antrieb [Online]. https://veekim.com/, Accessed 6 May 2024.
6. H.-U. Zabel, A model of human behaviour for sustainability. Int. J. Soc. Econ. **32**(8), 717–734 (2005)
7. D. A. Notter, R. Meyer, H.-J. Althaus, The western lifestyle and its long way to sustainability. Environ. Sci. Technol. **47**(9), 4014–4021 (2 April 2013)
8. J. Henrich, R. Boyd, S. Bowles, C. Camerer, E. Fehr, H. Gintis, *Foundations of human sociality: economic experiments and ethnographic evidence from fifteen small-scale societies* (Oxford University Press, Oxford, 2004)
9. M. Lewis, Signs you suffer from short-term thinking—and what to do about it (15 February 2024)
10. EY, How can boards convert sustainability from a wish to a winning reality? EY Europe Long-Term Value and Corporate Governance Survey (2024).
11. V. Waldersee, Mercedes-Benz delays electrification goal, beefs up combustion engine line-up. Reuters, [Online]. https://www.reuters.com/business/autos-transportation/mercedes-benz-hits-cars-returns-forecast-inflation-supply-chain-costs-bite-2024-02-22/. Accessed 7 May 2024.
12. Sceptical Science, Mason, 5 June 2020. [Online]. https://skepticalscience.com/history-climate-science.html.
13. American Fuel & Petrochemical Manufacturers, Don't ban our cars!, [Online]. https://www.dontbanourcars.com/about-afpm. Accessed 7 May 2024.
14. LSE (London School of Economics and Political Science), Daily Mail admits making up story about electric vehicles causing potholes (24 April 2024)
15. J. Morris, They are trying to kill the electric car again (10 February 2024), https://www.forbes.com/sites/jamesmorris/2024/02/10/they-are-trying-to-kill-the-electric-car-again/.
16. A. Winston, Why business leaders must resist the anti-ESG Movement, *Harward Business Review—Corporate social responsibility* (5 April 2023)
17. Factcheck: 21 misleading myths about electric vehicles, 2023 October 24. [Online]. Available: https://www.carbonbrief.org/factcheck-21-misleading-myths-about-electric-vehicles/. Accessed 2024 May 7
18. J. T. Jalles, Financial crises and climate change. Compar. Econ. Stud., pp. 1–25 (2023).
19. H. Ritchie, Many countries have decoupled economic growth from CO_2 emissions, even if we take offshored production into account, 2021 [Online]. https://ourworldindata.org/co2-gdp-decoupling
20. International Energy Agency, CO_2 Emissions in 2023—a new record high, but is there light at the end of the tunnel? (Paris, 2024)
21. H. Ritchie, M. Roser, Outdoor air pollution, 2022. [Online]. https://ourworldindata.org/outdoor-air-pollution
22. S. Dattani, L. Rodés-Guirao, H. Ritchie, E. Ortiz-Ospina, M. Roser, Life Expectancy, 2023. [Online]. https://ourworldindata.org/life-expectancy
23. M. Moatsos, Share of population living in extreme poverty, 2021. [Online]. https://ourworldindata.org/grapher/share-of-population-living-in-extreme-poverty-cost-of-basic-needs

Glossary

Activity variable (in Social Life Cycle Assessments [S-LCA]) Activity variables are a measure of an activity/process that is related to the activity/process output and can be scaled by the activity. An example is the number of working hours.

Albedo Proportion of the incoming light or radiation that is reflected by a surface. Dark colors have an albedo close to zero, meaning little or no energy is reflected. Pale colors have an albedo close to 100%, meaning nearly all the incoming energy (e.g. of the sun) is reflected. The albedo of the earth has an impact on climate change.

Area of Protection (in LCA) Different impact categories can be allocated to specific end-points that are termed Areas of Protection. These include human health/welfare, man-made structures and ecosystems, resources, natural environment/life-support functions of natural systems.

Attributional LCA Models the system linearly. The flows all increase linearly with the functional unit. It is a static model based on contemporary average data.

Backstop technologies or resources If resources become rare and expensive, alternatives get comparatively cheap, substituting the need for the scarce resource. For example, barium ferrite is a potential backstop resource for rare earth materials in electric magnets.

Biodiversity Variability among living organisms from all sources including but not limited to terrestrial, marine, and other aquatic ecosystems and the ecological complexes of which they are a part of. This includes diversity within species, between species, and of ecosystems.

Break-even point (originally in economics but transferred to other areas) Point at which the total costs and revenues are equal. In comparative life cycle studies, the point where the compared systems have the same results even if assumptions are changed.

Carbon neutral (according to ISO14068:2023) Â condition in which, during a specified period of time, the carbon footprint has been reduced as a result of greenhouse gas (GHG) emission reductions or GHG removal enhancements and, if greater than zero, is then counterbalanced by offsetting. In that there is a clear

© The Editor(s) (if applicable) and The Author(s), under exclusive license to Springer Nature Switzerland AG 2024
W. Schmidt, *Solutions For Sustainability Challenges*,
https://doi.org/10.1007/978-3-031-63624-0

hierarchy: Carbon neutrality is primarily achieved through GHG emission reductions, then GHG removal enhancements within the boundary of the subject, before offsetting.

Carbon footprint (according to ISO14068:2023) The sum of greenhouse gas (GHG) emissions and GHG removals of the subject (product, organization, etc.) expressed as carbon dioxide equivalents.

Carbon lock-in Equilibrium in which certain carbon-intensive technological systems (e.g. oil and gas extraction projects) or practices (e.g. structural state aid or behavior) persist over time, either "locking out" the potential for change or reducing further development towards low or net zero carbon emissions. This equilibrium directly or indirectly hinders the deployment of low-carbon technologies, which either cost more upfront or cannot compete with the lower prices of high-carbon technologies.

Characterization (in LCAs) Step in the Life Cycle Impact Assessment that determines the contribution of the elementary flows to each environmental impact category. For example, the impact category climate change is measured by the impact indicator Global Warming Potential (GWP) in terms of CO_2-equivalency. The characterization factor converts emissions into a common unit by considering their differences in efficacy in causing global warming: If 1 kg of methane has the same efficacy or potential for global warming as 28 kg of CO_2, the characterization factor of methane is 28. Elementary flows of CO_2 and methane are translated to their global warming potential using the different characterization factors of each gas.

Chemical recycling Process of converting polymeric waste by changing its chemical structure and turning it back into substances that can be used as raw materials for the manufacture of plastics or other products. Chemical recycling splits polymer chains and supplies products such as crude oil, naphtha, or fuels.

Circular economy System where materials never become waste and are kept in circulation through processes like maintenance, reuse, refurbishment, remanufacture, recycling, and composting or through regeneration of nature.

Classification (in LCA) Step in Life Cycle Impact Assessment in which elementary flows are assigned to the defined impact categories.

Consequential LCA Assessment of the system-wide change in pollution and resource flows that will result from a change in the level of the functional units produced. In doing so, consequential LCAs look in particular at those processes that are required incrementally to support this change. Therefore, consequential LCAs do not use static industry-average data but data representative of those processes used to satisfy the increased demand triggered by the incremental volumes produced.

Conventional LCC Reflects the cost for the producer or—as "cost of ownership"—the consumer.

Cradle-to-Cradle Trademarked term for an approach that is described as biomimetic where products/materials at their end-of-life are returned as "nutrients" into either technical or biological systems.

Cradle-to-gate Description of system boundaries that covers all processes from raw material extraction to final processes of production.

Cradle-to-grave Description of system boundaries that covers all processes from raw material extraction to final end-of-life processes of products.

Critical raw materials Raw materials of high economic importance, with a high risk of supply disruption due to their concentration of sources and lack of good, affordable substitutes.

Design thinking A mindset and an approach for developing innovative and life-centered solutions to complex problems in collaborative teams. It establishes an agile learning and working culture to thrive, change, and be sustainable in periods of digital transformation. In practical terms, design thinking is an iterative process to understand users/consumers, challenge assumptions, redefine problems, and create innovative solutions which are first prototyped and then tested. The overall goal is to identify alternative strategies and solutions that are not apparent immediately with the initial level of understanding.

Discounting Process of converting a value received in a future time period to an equivalent value received immediately. Discounting represents the time preference of when costs and benefits occur. It accounts for the time preference/value of money by converting all costs and profits over the life cycle time into their present value.

Do no significant harm Precautionary principle in various sustainable finance regulations like the EU taxonomy with the intention of avoiding economic activities that are detrimental to the environment. In EU taxonomy, this refers to the six areas (1) climate change mitigation, (2) climate change adaptation, (3) sustainable use and protection of water and marine resources, (4) transition to a circular economy, (5) pollution prevention and control, and (6) protection and restoration of biodiversity and ecosystems).

Down-cycling Subjective term that suggests a specific recycling approach leads to lower quality product.

Earth system Earth's interacting physical, chemical, and biological processes on land, in oceans, in the atmosphere and at the poles. It includes the planet's natural cycles—the carbon, water, nitrogen, phosphorus, sulfur, and other cycles—and deep Earth processes as well as life on Earth.

Ecological footprint See environmental footprint.

Economies of scale Cost and efficiency advantages companies experience when production is scaled up as capital, overhead, and other costs can be spread over a larger amount of goods.

El Niño Natural climate phenomenon marked by warmer-than-average sea surface temperatures in the central and eastern Pacific Ocean near the equator, which occurs on average every 2–7 years. El Niño's impact on the climate extends far beyond the Pacific Ocean.

Elementary flow (in Life Cycle Assessments) All materials, substances, energies, spaces etc. that are directly taken from the environment or released directly back into the environment: all input and output flows between the system and the environment.

End-point quantifiable cause-effect chain (in Life Cycle Assessment) Methodology for modeling the impact categories. It follows the cause-effect chain from elementary flows to the final damages caused to an Area of Protection such as human health, biodiversity or natural resources for environmental LCAs or well-being of stakeholders, human rights, etc. for social LCAs.

End-of-life Processes after the use of a product including collection, processing, recycling, recovery, and final disposal.

Environmental footprint Measure for the impact of an activity on the environment based on holistic life cycle thinking. Types of environmental footprints include carbon, water, resource, atmospheric, waste footprint, or footprint on biodiversity. The EU regulates environmental footprints for products.

Environmental LCC Life Cycle Costing with system boundaries and functional units equivalent to environmental LCAs. Environmental LCCs cover internal costs and those external costs (externalities) that are expected to be internalized unless already covered in Life Cycle Assessments.

Externalities (in the context of this book) Indirect effects of an economic activity on third parties that are not included in the price of the economic activity.

Foreground process (in Life Cycle Assessments) Processes that are related to the considered product/service. Specific data is often collected for these processes.

Full LCA Life Cycle Assessments with a full scope, i.e. covering all elementary flows from raw materials extraction to final end-of-life operations according to ISO 14040/14044:2006.

Functional unit A quantified description of the performance requirements that the product system fulfills in Life Cycle Assessments.

Futuring Envisioning and planning for potential futures, often through a systematic process that includes analysis, exploration, and forecasting.

Gig economy Labor market that relies heavily on temporary and part-time positions filled by independent contractors and freelancers rather than full-time permanent employees.

Greenwashing Misrepresentation, misstatement, and false or misleading practices in relation to environmental, social, and governance credentials.

Hot spot (in sustainability assessments) Area or life cycle stage with the most important impacts on sustainability.

Impact category (LCA, S-LCA) Areas of environmental or social concerns like climate change or forced labor.

Impact category indicator (LCA, S-LCA) Quantitative representation of the impact category, linking elementary flows to the impact category using characterization factors.

Industry 4.0 and 5.0 Industry 4.0 is the digitalization of manufacturing based on technologies such as the Internet of Things and big data. Smart factories are equipped with advanced sensors, embedded software, and robotics that collect and analyze data in real time. Industry 5.0 seeks to add human, environmental, and social aspects back into the equation, i.e. refers to robot and smart machines working alongside people, adding resilience including sustainability goals into industry.

Internet of Things The collective network or interconnection via the internet of computing or connected devices embedded in everyday objects, enabling them to send and receive data. It includes the technology that facilitates communication between devices and clouds, as well as between the devices themselves.

Life Cycle Process steps including raw material extraction, material, part and product production, use of the product, and final end-of-life processes.

Life Cycle Impact Assessment Third step in Life Cycle Assessment where the Life Cycle Inventory results of all elementary flows (resources, emissions, etc.) are assessed for their environmental impact.

Limits to Growth (in report of Club of Rome) Suggestion that the global system of nature cannot support present rates of economic and population growth beyond the year 2100 even with advanced technology.

LOHAS Abbreviation for the consumer group "Lifestyle of Health and Sustainability". Aspects of sustainability such as environmental protection, social justice and health are among the main issues in the lifestyle and consumption of LOHAS. Despite the focus on products with perceived better environmental and social footprint, factors such as entertainment and fun remain important for LOHAS. Customers who are LOHAS are considered critical, creative, progressive, curious, and idealistic. And these people often have a high income, which they are happy to invest in perceived sustainable products.

LOVOS Abbreviation for the consumer group "Lifestyle of Voluntary Simplicity". LOVOS follow voluntary (or not voluntary) a "less is more" lifestyle and consumption pattern. The aim is to live a highly simplified lifestyle in which consumption is completely renounced. LOVOS want to escape the constraints of everyday life and lead as self-determined a life as possible.

Materiality Identification of issues that matter most for the organization, the life cycle, the study, etc.

Mechanical recycling Recycling that preserves the molecular structure. It mechanically crushes the material and in the case of thermoplastics remelts it into granulate. This granulate is then used to make new plastic products.

Midpoint Impact assessment approach in LCAs or S-LCAs. It follows the cause-effect chain from elementary flows to changes in the natural environment without capturing the full consequences for the final "Areas of Protection" (end-points). An example of a midpoint indicator is the Global Warming Potential (GWP) expressed in CO_2 equivalent emissions.

Net Present Value (NPV) Translation of the future value of an investment into today's value assuming a certain discounting rate. The resulting present value is compared to the necessary investment.

Net zero Condition in which anthropogenic GHG emissions are balanced by anthropogenic GHG removals over a specified period. Term generally applied to territories (e.g. a country, a municipality) and to organizations, but not to products.

Planetary boundaries The set of nine planetary boundaries within which humanity can continue to develop and thrive for generations to come. The nine considered areas are biosphere integrity, climate change, stratospheric ozone depletion, ocean acidification, biogeochemical flows, land system change, freshwater change, atmospheric aerosol loading, novel entries. They reflect nine processes that are critical for maintaining the stability and resilience of Earth's systems as a whole. The methodology is based on earth science modeling with the target of maintaining the conditions of the Holocene i.e. the current geological epoch.

Platform work Any work organized through a digital labor platform on the basis of a contractual relationship between the digital labor platform or an intermediary and the individual, irrespective of whether a contractual relationship exists between the individual or an intermediary and the recipient of the service.

Polluter-pays principle Producers of pollution should bear the costs of managing them in terms of preventing damage to human health or the environment.

Post-consumer recyclates (PCR) Materials originating from waste materials that are passed through the hands of consumers. These waste materials are generated by households or by commercial, industrial, and institutional entities as products that can no longer be used for their intended purpose. It also includes the return of materials from the distribution chain.

Post-industrial recyclates (PIR) or pre-consumer recyclates Materials that are diverted from the waste stream during a manufacturing process. This excludes the reutilization of materials through processes such as reworking, regrinding, or scraping. Materials can then be reclaimed within the same process that generated them. These waste streams must have been subjected to at least one additional process step.

Product Environmental Footprint Type of regulated Life Cycle Assessment for a product.

Recycled materials Materials originating from the collection and processing of waste materials.

Reference flow Measurement of the required inputs into a process in a system that is needed to fulfill the functionality of the defined functional unit.

Refurbish Involves restoring used technology to a cosmetically close-enough-to-new state and fixing any minor issues. Professional technicians refurbish products to grant them a factory-new performance with different grades of refurbishment determined by extensive testing and quality controls. Refurbishment can range from a thorough check for defects to a complete cleanup and rebuild of the product to ensure a perfect working condition.

Remanufacturing Remanufacturing is the process of rebuilding a product to meet or exceed the original manufacturer's specifications using a combination of reused, repaired, and new parts. It may show a level of wear, but in terms of quality and robustness, a remanufactured part is equivalent to a brand-new part. It is typically sold with a full warranty.

Renewable materials Materials derived from plant, animal, or bacterial resources that are sustainably cultivated and regenerated within a human lifetime (like hemp, flax, kenaf, jute, wood, and other).

Return-on-Invest Profit made as a result of an investment. It is calculated typically as the ratio between the net profit over the total amount of the investment within a certain period of time.

Scope 1 carbon emissions Greenhouse gas emissions from sources that an organization owns or controls directly such as emissions from an organization's own manufacturing process (for a manufacturer) or car emissions for a service organization owning an own car fleet.

Scope 2 carbon emissions Greenhouse gas emissions from an organization's purchased energy such as CO_2 emissions from the generation of electricity supplied to the organization.

Scope 3 carbon emissions Greenhouse gas emissions under indirect responsibility of an organization up and down the life cycle/value chain that are not covered under scope 1 or 2 such as car emissions of vehicles sold for an automotive manufacturer or the emissions of paper production for a newspaper publisher.

Scope 4 carbon emissions The reduction in emissions achieved by using a company's products or service.

Social handprint Positive social impacts of an organization or activity. Handprints are (1) a measure of positive action, collaboration, and networking towards sustainability and a tool for measuring the positive impact of actions in promoting sustainable development, (2) a normative approach (what should be done and not only what has been done), (3) a measure of action by individuals who support measurable changes in behavior towards sustainable development and the environment [8], (4) actions to improve the well-being of people or the sustainability or healing of the planet and (5) focused on positive attitudes to sustainability and taking appropriate action.

Societal LCC Life Cycle Costing holistically considering all costs linked to an economic activity from a societal perspective (rather than only a producer or customer perspective). Therefore, transfer payments within the society—such as subsidies and taxes—are excluded because they are internal to the system.

Sustainable Development Goals (SDGs) United Nations' 17 Sustainable Development Goals for 2030 like no poverty, reduced inequalities, and climate action.

Technical Sustainability Management Combination of sustainable management, good governance, and engineering practices based on life cycle thinking that improves the relevant social, environmental, and economic aspects of products and organizations holistically.

Tipping point Critical threshold beyond which a system reorganizes, often abruptly and/or irreversibly.

Total Cost of Ownership Cost analysis that aims to uncover all the likely costs that result from owning assets during ownership life. It attempts to reveal and summarize ownership costs that are obvious and known before purchase, but also cover so-called "hidden costs" that are easy to overlook.

Trade-off A trade-off is when you choose one thing which causes you to have to give up, or sacrifice, another.

Value Impact Quantification and valuing of a company's positive and negative externalities related to society and the environment.

Well-to-wheel (WtW) Analysis of greenhouse gas emissions and energy consumption for transport vehicles with system boundaries that include all processes required to produce the fuel/energy together with the emissions from the consumption of these fuels/energy during the use of the vehicle. Excludes processes for the production and end-of-life of vehicles.

Well-to-tank (WtT) Analysis of greenhouse gas emissions and energy consumption for transport fuel/energy with system boundaries that include all processes required to produce the fuel/energy. Excludes processes for the production and end-of-life of vehicles as well as the emissions during the use of vehicles.

Tank-to-wheel (TtW) Analysis of greenhouse gas emissions and energy consumption for transport vehicles with system boundaries that only include the emissions from the consumption of the fuels/energy during the use of the vehicle. Excluded are all processes for the production and end-of-life of vehicles as well as the production of fuels/energy.

Up-time Time during which a machine or a vehicle is fully operable/not in maintenance or under repair.

Willingness to pay The maximum price that a customer is willing to pay for a product, service, or action.

SPRINGER NATURE

GPSR Compliance

The European Union's (EU) General Product Safety Regulation (GPSR) is a set of rules that requires consumer products to be safe and our obligations to ensure this.

If you have any concerns about our products, you can contact us on ProductSafety@springernature.com

In case Publisher is established outside the EU, the EU authorized representative is:

Springer Nature Customer Service Center GmbH
Europaplatz 3
69115 Heidelberg, Germany

The manufacturer's authorised representative in the EU is Springer Nature Customer Service Centre GmbH, Europaplatz 3, 69115 Heidelberg, Germany. If you have any concerns regarding our products, please contact ProductSafety@springernature.com

Printed and bound by CPI Group (UK) Ltd, Croydon, CR0 4YY
01/10/2026
02225969-0010